Analytical Evaluation of Uncertainty Propagation for Probabilistic Design Optimisation

Online at: https://doi.org/10.1088/978-0-7503-4931-4

Analytical Evaluation of Uncertainty Propagation for Probabilistic Design Optimisation

Melanie Po-Leen Ooi
School of Engineering, University of Waikato, Hamilton, New Zealand

Arvind Rajan
Clear21, Melbourne, Australia

Ye Chow Kuang
School of Engineering, University of Waikato, Hamilton, New Zealand

Serge Demidenko
School of Food and Advanced Technology, Massey University, Auckland, New Zealand
and
School of Engineering and Technology, Sunway University, Sunway City, Malaysia

IOP Publishing, Bristol, UK

ISBN 978-0-7503-4931-4 (ebook)
ISBN 978-0-7503-4929-1 (print)
ISBN 978-0-7503-4932-1 (myPrint)
ISBN 978-0-7503-4930-7 (mobi)

DOI 10.1088/978-0-7503-4931-4

Version: 20230601

IOP ebooks

British Library Cataloguing-in-Publication Data: A catalogue record for this book is available from the British Library.

Published by IOP Publishing, wholly owned by The Institute of Physics, London

IOP Publishing, No.2 The Distillery, Glassfields, Avon Street, Bristol, BS2 0GR, UK

US Office: IOP Publishing, Inc., 190 North Independence Mall West, Suite 601, Philadelphia, PA 19106, USA

Contents

Acknowledgements

The authors would like to extend their gratitude to their research collaborators from different parts of the world.

First, they would like to acknowledge Herman Carstens from Invoke Analytics for his invaluable contribution and support. The authors are also grateful to Sarbajit Paul from Dong-A University and Professor Chang, who provided them with the opportunity to work with their team and learn from their expertise, which led to significant improvement in the probabilistic design optimisation framework. The authors also express their appreciation to the teams of Professor (Barry) Yu Bai and Fujia Luo, and Associate Professors Akhil and Ankit Garg from Shantou University, as well as Dr Venkatesh from UNSW for their collaboration and support in the real-world case studies.

The authors also thank Sorena Artin from Asia Pacific International College and Alexander Andreychenko from Saarland University for sharing their knowledge and expertise. The authors are grateful to all the administrative and technical staff from the School of Engineering at Monash University, Malaysia, especially Ms Krisnakumari Perumal, Ms Nurul Rukiah, Ms Misha, Mr Manan, Mr Suresh, Mr Hasnan, Ms Rozidah, Mr Daniel, Ms Irdawani, Ms Laura, and Ms Shamini. The valuable input of Dr Houshyar Pajooh of Sunway University to the specialist online software tools redevelopment is appreciated.

The authors would also like to acknowledge the grant-making bodies who have supported this work, including the Malaysian Ministry of Education for their Fundamental Research Grant Support and the IEEE Instrumentation and Measurement Society for the Graduate Fellowship Award.

Author biographies

Melanie Po-Leen Ooi

Melanie Po-Leen Ooi received her BEng (Hons), MEngSc (Research), and PhD from Monash University in the areas of electronics, computer engineering, and computational intelligence. She is currently an Associate Professor and Assistant Dean (Research) of the School of Engineering, University of Waikato, New Zealand as well as an Adjunct Professor of the School of Engineering and Technology, Sunway University, Malaysia. She is the recipient of several awards including the 2019 Rutherford Discovery Fellowship from the New Zealand Royal Society, the 2017 Mike Sargeant Career Achievement Award from the Institution of Engineering and Technology (IET), the 2014 Outstanding Young Engineer Award from the Instrumentation and Measurement Society (I&MS) of the Institute of Electrical and Electronics Engineers (IEEE), 2014 Excellence Award from the International Education Association of Australia, 2011 Citation for Outstanding Contributions to Student Learning from the Australian Learning and Teaching Council. She was the youngest female elevated to the IET Fellow grade. Associate Professor Ooi is a UK Chartered Engineer and Senior Member of the IEEE. She is a guest editor for the *IEEE Transactions in Instrumentation and Measurement*, an Administrative Committee Member of IEEE I&MS as well as a Secretary of the Society's Technical Committee on Fault-Tolerant Measurement Systems - TC-32.

Arvind Rajan

Arvind Rajan is a Data Scientist at Clear21 in Melbourne, Australia. He received his BEng (First Class Hons) and PhD degrees from Monash University, in 2015 and 2018, respectively. He is a Member of IEEE, IEEE I&MS, and its Technical Committee on Fault-Tolerant Measurement Systems – TC-32. He is also a Member of IET. He was the winner of the IEEE I&MS 2017 Graduate Fellowship Award and was awarded the Monash University 2018 Vice-Chancellor's Commendation for Doctoral Thesis Excellence.

Ye Chow Kuang

Ye Chow Kuang received his BEng (Hons) in electromechanical engineering and PhD in non-invasive diagnostic techniques from the University of Southampton. From 2005 to 2018 he was with Monash University specialising in machine intelligence, machine vision, and uncertainty modelling in engineering design. He is currently a Senior Lecturer with the Faculty of Science and Engineering, University of Waikato, New Zealand. He is Chair of the Technical Committee on Fault-Tolerant Measurement Systems of the IEEE I&MS - TC-32. He is also a Member of IET and UK Chartered Engineer. He was a co-recipient of the 2014 Excellence Award from the International Education Association of Australia, and the 2012 IEEE I&MS Faculty Course Development Award.

Serge Demidenko

Serge Demidenko received a five-year electrical engineer qualification in computer engineering from the Belarusian State University of Informatics and Radio Electronics, and PhD from the Institute of Engineering Cybernetics of the Belarusian Academy of Sciences. He is currently a Professor at the School of Food and Advanced Technology, Massey University, New Zealand. He is also an Adjunct Professor at Sunway University, Malaysia where he previously was a Dean at the School of Engineering and Technology. Professor Demidenko is a Chair of the IEEE Technical Field Awards Council, Member of the IEEE Awards Board, and Past Chair of the IEEE Joseph F. Keithley Award in Instrumentation and Measurement. He is a Fellow of IEEE and IET, UK Chartered Engineer as well as a Member of the European Academy of Sciences and Arts.

Nomenclature

Greek symbols

β	Reliability index
χ	Matrix condition number
δ	Scale parameter
γ	Lagrange multiplier for orthogonal basis polynomials
μ	Mean
Ω	Skewness
ω	Shape parameter
Φ	CDF of the standard normal distribution
ϕ	Maximal mass
Ψ	Kurtosis
Ψ_e	Excess kurtosis
ρ	CDF bound gap
σ	Standard deviation
ζ_d	Design window constant
ζ_s	Surface region constant
τ	Location parameter
ε	Coverage factor estimation error
ε_{10}	ε below 10% percentile range
ε_{90}	ε above 90% percentile range
ε_{max}	Maximum of ε_{10} and ε_{90}
φ	Lagrange multiplier

Other symbols

d	Design variable
$\mathbf{d}$	Vector of di for $i = 1, \dots, N_d$
$\mathbf{d}^L$	Vector of design lower bounds
$\mathbf{d}^U$	Vector of design upper bounds
M	MC sample size
N_c	Number of constraint functions
$\widetilde{m}$	Preconditioned moments
N_d	Number of design variables
N_m	Number of moments
N_q	Number of quadrature points
N_r	Number of random input variables
$\widetilde{\varphi}$	Lagrange multiplier corresponding to $\widetilde{m}$
P	Multivariate polynomial
$\mathbb{R}$	Set of real numbers
S	Shannon entropy
V	Random variable with standard distribution
v	Realisation of v
X	Random input variable
x	Realisation of X
$\mathbf{X}$	Vector of X_i for $i = 1, \dots, N_r$
$\mathbf{x}$	Vector of realisations of $\mathbf{X}$
Y	Random output variable
y	Realisation of Y

$\mathbb{Z}$	Set of integers
$\mathbb{Z}^-$	Set of negative integers
$\mathbb{Z}^+$	Set of positive integers

Acronyms/abbreviations

AMV	Advanced mean value
ANUNCEATOR	Analytical uncertainty calculator
CDF	Cumulative distribution function
CF	Cornish–Fisher
CMV	Conjugate mean value
CPU	Central processing unit
DRM	Dimension reduction method
EGLD	Extended generalised lambda distribution
FE	Finite element
FEV	Function evaluation
FORM	First-order reliability method
GA	Genetic algorithm
GBD	Generalised beta distribution
GLD	Generalised lambda distribution
GOPoly	Generalised orthogonal polynomial
GUM	Guide to the expression of uncertainty in measurement
HMV	Hybrid mean value
IEEE	Institute of Electrical and Electronics Engineers
IF	Impact factor
IMS	Instrumentation and Measurement Society
ISO	International Organization for Standardization
JCR	Journal Citation Reports
MaxEnt	Maximum entropy
MBIR	Moment-based integration rule
MC	Monte Carlo
MPP	Most probable point
M&V	Measurement and verification
PDF	Probability distribution function
PFD	Peak flux density
PMA	Performance measure approach
PMI	Performance moment integration
PMOR	Parametric model order reduction
PolyMoment	Polynomial moment
RAM	Random-access memory
RBDO	Reliability-based design optimisation
RBRDO	Reliability-based robust design optimisation
RDO	Robust design optimisation
RIA	Reliability index approach
RSM	Response surface methodology
SLA	Single-loop approach
SLS	Serviceability limit state
SORA	Sequential optimisation and reliability assessment
SORM	Second-order reliability method
SVM	Support vector machine
TC	Technical Committee

TGH	Tukey's *gh*
THD	Total harmonic distortion
ULS	Ultimate limit state
XGLD	Extended generalised lambda distribution

Chapter 1

Introduction

The use of sensor systems in combination with machine learning and computational science to make data-driven decisions is increasing at an unprecedented rate. The challenge has shifted from simply acquiring data to acquiring high-quality, valuable data. Not all data are equally valuable, and this is especially true for sensor data, which can be flawed and have a low signal-to-noise ratio. If high- and low-quality data are processed and then used jointly and equally in decision-making algorithms, it can result in incorrect and potentially harmful conclusions for infrastructure and mission-critical applications, such as building design, that can impact lives, the environment, and the economy.

One important way to assess the value of data is to conduct uncertainty analysis, which aims to determine the range of possible values for a physical quantity's true value. This practice has been given great importance for many decades by professionals such as metrologists, instrumentation engineers, and measurement device practitioners. In recent years, however, it has become increasingly important for designers in various engineering fields to produce high-quality design solutions for complex engineering systems. Chapter 2 of this book examines some of the methods used for uncertainty evaluation.

Incorporating uncertainties into the design process and solution for a complex system is a challenging task, as the relationship between the output of interest and the input variables is often unclear. A well-designed technical system should not only operate effectively, but also be the most resource-efficient solution that complies with its design constraints. This is a difficult optimisation problem, as it requires a set of parameters to be found that optimises the system's performance while satisfying conflicting constraints, such as sustainability and affordability. Additionally, each system parameter comes with its own uncertainties. Within the domain of probabilistic design optimisation, the uncertainties are incorporated by including an additional procedure of reliability and/or robustness analyses into the

doi:10.1088/978-0-7503-4931-4ch1

framework. There are several approaches to achieving this. Chapter 3 covers the general methods and formulations.

The authors of this monograph believe that it is important to go a step further and incorporate the propagation of uncertainty into the probabilistic design optimisation process.

The statistical properties of any given system input are generally known (through incoming signal measurements, manufacturer specifications, *a priori* assumptions, etc). The known uncertainties at the input then propagate through the system, allowing the derivation of probability density functions or cumulative distribution functions of the design constraints imposed on the system, i.e. to characterise the uncertainties. This, in turn, facilitates the system analysis (if and where it is needed).

Mathematical expressions are used to model the world through conceptual representations of physical quantities together with their corresponding measurement methods and scales. Using mathematical operations, the interactions between different quantities are numerically encoded. These models form the foundation upon which our understanding of the physical world is built. They allow scientists and engineers to predict what would happen in the real world if they were to design and deploy a particular system.

Model equations are extremely important in the field of measurement. While some physical quantities can be directly measured, others are inferred by applying the appropriate model equations together with measurements of other physical quantities. A common example is the radar speed trap, whereby the speed of a vehicle, as described in equation (1.1), is inferred from the difference between the transmitted and received frequencies by exploiting knowledge of the Doppler effect. In this example, speed is not directly measured; instead, it is inferred from the frequency of the transmitted radio wave f_0 and the change in frequencies Δf and then related to the vehicle speed v via knowledge of the angle between the radar and the road ϕ, the refractive index of air n, and the speed of light in vacuum c:

$$\text{Speed of vehicle,}\quad v = \frac{\Delta f\,.\,c}{2nf_0\cos(\phi)}. \tag{1.1}$$

There are thousands of mathematical models or equations available today that can be used to infer unmeasurable or unobservable quantities from observable ones. To generalise: a chosen measurement equation $Y = g(X_1, \ldots, X_N)$ describes the indirect measurement of a quantity of interest Y and its relationship to the measurement of known quantities $X_1, \ldots, X_N$. This equation can then be expanded to include the effect of measurement error in the measured quantities, leading to an expression known as the *propagation of error*. This relates the error in the indirect measurement of the quantity of interest to the errors of the contributory measurements in the equation. This is also known as the *propagation of uncertainty*.

The Guide to the Expression of Uncertainty in Measurement (GUM) [1] published by the *Bureau International des Poids et Mesures* (BIPM) is one of the most widely referenced documents in the field of uncertainty evaluation and uncertainty propagation. It outlines the above measurement problem alongside the general rules for the expression and evaluation of measurement uncertainty that can be applied to a wide range of industrial and scientific measurements. Here, the quantity of interest Y or the 'quantity intended to be measured' is referred to as the measurand [2].

An addendum to the GUM was presented in 2020 that outlined a process for developing and using measurement models. This addendum was adopted a year later by the ISO/IEC [3]. This addendum contains a specific section on the representation of polynomials. The work outlined in this monograph follows the GUM framework closely in that it:

1. Examines the estimators that might be applicable for extending the basic measurement model used to appraise the measurand using the measured quantity values of the inputs
2. Provides guidance on the determination of probability distributions for the quantities appearing in the measurement model
3. Represents the measurement model by means of polynomials to facilitate an analytical determination of the uncertainty
4. Utilises the 'new' measurement model and its corresponding analytical uncertainty expressions within probabilistic design optimisation procedures

Two methods are commonly used to estimate the uncertainty of a measurement; these are known as as type A and type B evaluations. The GUM framework, which is closely followed in the addendum to the GUM presented in 2020, highlights the importance of estimating measurement uncertainty in order to ensure the accuracy and reliability of industrial and scientific measurements. Within the framework, type A and type B evaluations are used to estimate the uncertainty of a measurement based on different sources of information.

Type A evaluations of uncertainty are based on the statistical analysis of repeated measurements, which are used to estimate the variability of the measurement. This type of evaluation is used to estimate the random error component of the uncertainty and is considered to be a good representation of the uncertainty when the measurements are independent and identically distributed. On the other hand, type B evaluations of uncertainty are based on information that is not directly related to the measurement itself. This can include expert judgments, published data, or data obtained from other sources. Type B evaluations of uncertainty are used to estimate the systematic error component of the uncertainty, which is considered to be a good representation of the uncertainty when the measurement process is not well understood. Together, type A and type B evaluations of uncertainty provide a comprehensive estimate of the total uncertainty of a measurement that can be used to assess the accuracy and reliability of the measurement results.

The essential fundamentals in the field of uncertainty are covered in chapter 3, which outlines those formulated in the GUM. The immediate and direct practical application of the GUM for the design optimisation of complex systems is often problematic, if not impossible. Thus, chapters 4 and 5 address the challenge by introducing a moment-based analytical approach for the evaluation of standard and expanded uncertainties, respectively. This technique is then applied within the

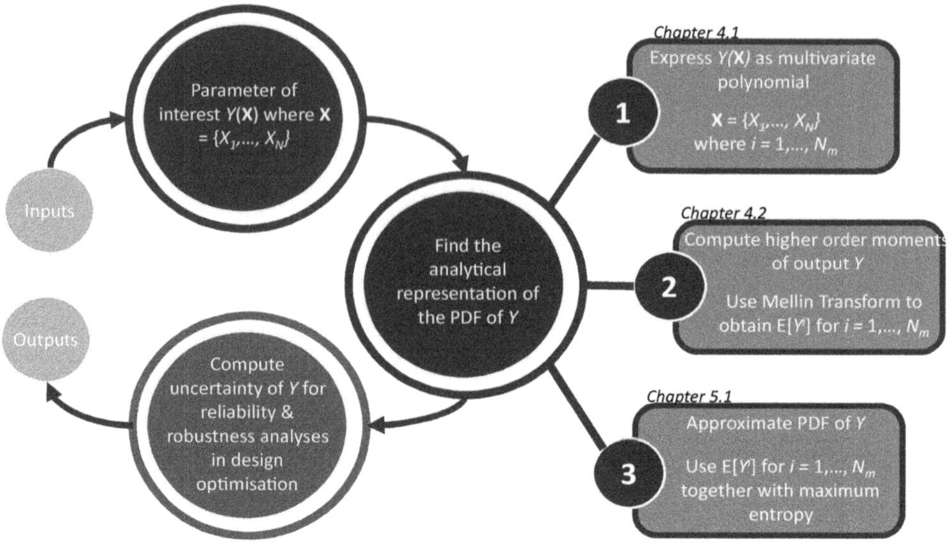

Figure 1.1. The application of moment-based uncertainty evaluation in reliability and robustness analysis for design optimisation (chapters 4 and 5).

probabilistic design optimisation framework, specifically *reliability-based design optimisation* (RBDO), *robust design optimisation* (RDO) and *reliability-based robust design optimisation* (RBRDO), as shown in figure 1.1. The application-related aspects of the developed probabilistic design optimisation framework are presented alongside real-world case studies in chapter 6.

References

[1] Guides in Metrology Committee: JCGM 2008 *Evaluation of measurement data—guide to the expression of uncertainty in measurement* (Sèvres: BIPM) https://www.bipm.org/en/committees/jc/jcgm/publications JCGM 100:2008

[2] Guides in Metrology Committee: JCGM 2012 *International vocabulary of metrology—basic and general concepts and associated terms (VIM)* 3 (Sèvres: BIPM) https://www.bipm.org/en/committees/jc/jcgm/publications JCGM 200:2012. This 3rd edition is also published as an ISO Guide.

[3] ISO/IEC Guide 98–6:2021 2021 *Uncertainty of Measurement—Part 6: Developing and Using Measurement Models* (International Organisation for Standardization)

Chapter 2

Uncertainty propagation

Assuming that a measurement model is known, the measurand can be estimated from the repeated measurement of input quantities. According to the 2020 Part 6 addendum to the GUM [1], a measurement model constitutes ‘a relationship between measurands and input quantities to be involved in the measurement’. It is always assumed in this monograph that a measurand cannot be directly measured and that there are some known theoretical or mathematical relationships between the inputs and the outputs.

Unfortunately, it is impractical or, in some cases, impossible to mathematically derive or model the distributions for each possible system output. In such cases, the distribution can be determined by measuring input quantities, so that the relationship between the input and output quantities is known. This technique enables output uncertainties to be inferred, given the known uncertainties of the input. The measurement model may therefore take a broader form than a theoretical relationship, as it may include an empirical relationship or a hybrid of the two. For example, the measurement model may require an extension to its mathematical model to incorporate temperature or pressure effects, and this now becomes the new measurement model that a design optimisation procedure should use.

The first subsection of this chapter summarises three methods of estimating a measurand. These can be used to extend the theoretical measurement model by evaluating the measurand uncertainty after performing measurements of the input. The four subsequent subsections summarise some of the methods for the assignment of probability distributions to the quantities appearing in the measurement model. These are: (1) first-order linearisation with normal approximation; (2) the *Monte Carlo* simulation method; (3) analytical methods that use mathematical representations of the *probability density function* (PDF); and (4) uncertainty evaluation using moments.

In an effort to minimise notational complexity, the discussions presented in the text denote all outputs as Y and their corresponding mathematical functions as $Y(\,\cdot\,)$

rather than $g(\cdot)$. The model equations are related to the known quantities $\mathbf{X} = \{X_1, \ldots, X_N\}$. It should be noted that the GUM [2] represents the input–output relationship as $Y = g(\boldsymbol{X})$, where $g(\cdot)$ denotes the mathematical function.

This chapter omits polynomial chaos theory [3, 4], which is a full analytic method that is based on the orthogonal decomposition of the Hilbert space generated by a probability measure [5]. However, this theory is not within the GUM framework. A review of the moment method is, however, comprehensively presented in this chapter, as it leads up to the discussions in chapter 4.

2.1 Averaging methods for estimating the measurand in nonlinear models

The GUM outlines two methods for estimating the measurand (the quantity being measured) in nonlinear models. Method 1 averages the estimates of the input quantities, while Method 2 averages the estimates of the measurand (represented as a distribution). Method 2 is preferred for nonlinear measurement functions and in cases in which the definitional uncertainty of the input quantities outweighs their sample acquisition uncertainty. A supplementary method, Method 2S, improves computational efficiency in specific cases. Method 2S requires the definitional uncertainty to outweigh the acquisition uncertainty and the measurement function to be known and modelled in a specific form.

The GUM states that an 'incomplete definition of the measurand can give rise to a component of uncertainty sufficiently large that it must be included in the evaluation of the uncertainty result'. Further in-depth analysis performed in [6, 7] shows that the effects of the 'incomplete definition of the measurand' that should be taken as a part of the measurand constitute the definitional uncertainty [8], which is also known as the 'intrinsic' uncertainty in the GUM Section D3.4 [2]. As a result, the measurand must be expressed as a statistical distribution and the measured quantity value should be interpreted as the expected value of the corresponding distribution [7]. On the contrary, if the definitional uncertainty is negligible and the measurand is only affected by some acquisitional uncertainty, then the measured quantity value can essentially be expressed as a unique value under certain assumptions, as discussed in [9, 10].

The GUM outlines two estimators that can be used when repeated independent measurements of the input quantities $X_1, \ldots, X_N$ are taken in order to estimate the measurand Y. They are known as *Method 1* and *Method 2* in the literature [6, 7, 9, 10] and are shown as $\hat{Y}$ and $\overline{Y}$ in equations (2.1) and (2.2), respectively.

$$\hat{Y} = g(\overline{X}_1, \ldots, \overline{X}_N), \quad \text{where } \overline{X}_i = \sum_{k}^{K} \frac{X_{i,k}}{K} \tag{2.1}$$

$$\overline{Y} = \frac{1}{K}\sum_{k=1}^{K} Y_k = \frac{1}{K}\sum_{k=1}^{K} g(X_{1,k}, X_{2,k}, \ldots, X_{N,k}), \tag{2.2}$$

where K is the number of observations (or sample size), $X_{i,k}$ is the kth independently observed value of the input quantity X_i for $i = 1, \ldots, N$, and N is the number of input quantities.

Method 1 averages the estimates of the input quantities to minimise the propagation of acquisition uncertainty through the measurement function, while Method 2 averages the estimates of the measurand (represented as a distribution) to find its expected value. The estimators $\hat{Y}$ and $\overline{Y}$ are generally nonequivalent, and section 4.1.1 of the GUM states that Method 2 'may be preferable' over Method 1 for nonlinear measurement functions. This is an important observation, since the focus of this monograph is the use of uncertainty in the design optimisation process of complex systems—i.e. those in which the measurement function could be highly nonlinear.

The investigation in [10] shows that there is no uniformly preferred method in all of the presented cases, while [6, 7] suggested that Method 2 is preferred whenever the definitional uncertainty of the input quantities outweighs the sample acquisition uncertainty and vice versa. Further work [11] provides a supplementary method to Method 2, known as *Method 2S*, shown in equations (2.3) and (2.4). It produces a mean square error that is smaller than that of Method 2 while improving the computational efficiency whenever the measurement function involves multiplicative operations between independent random variables.

The two conditions for the application of Method 2S are:

Condition 1: the definitional uncertainty of the measured input quantities outweighs the acquisition uncertainty

Condition 2: the measurement function is exactly known and is modelled using the sum-of-product form shown in equation (2.3):

$$Y = \sum_{r}^{R} \prod_{q}^{Q_r \leqslant N} \mathcal{P}_{qr}\left(X_{I_{qr}}\right), \tag{2.3}$$

whereby $\mathcal{P}_{qr}(\cdot)$ is the qth factor that makes up the rth summation term and $X_{I_{qr}}$ is its corresponding set of uncorrelated input quantities. The sets $X_{I_{qr}}$ with different q but the same summation term r are mutually exclusive subsets of the N input quantities $X_1, \ldots, X_N$, which are assumed to be statistically independent. Hence, the subscript I_{qr} refers to sets of positive integers that are mutually exclusive for the same r.

$$\check{Y} = \sum_{r}^{R} \prod_{q}^{Q_r \leqslant N} \left(\frac{1}{K}\sum_{k=1}^{K} \mathcal{P}_{qr}\left(X_{I_{qr},k}\right)\right) \tag{2.4}$$

Here $X_{I_{qr},k}$ denotes the kth independently observed values of $X_{I_{qr}}$.

The only difference between Method 2 and Method 2S is that the former performs the averaging of the independent product terms after the multiplication, while the latter does the averaging before the multiplication. The same paper [11] shows that Method 2S allows more efficient use of the available measurement data for a specific class of problem. This is especially useful in Type-A evaluations in which the uncertainty is determined by experimental (and therefore physical and/or financial) constraints that can often limit the number of samples.

2.2 First-order linearisation and the normality assumption

The first-order linearisation method is illustrated in figure 2.1. This is a procedure that finds the standard deviation of Y based on the first-order Taylor series approximation (or linearization) of a function $Y(\cdot)$, given N_r random input quantities $\mathbf{X} = \{X_1, \ldots, X_{N_r}\}$:

$$\sigma_Y = \sqrt{\sum_{i=1}^{N_r}\left[\frac{\partial Y}{\partial X_i}\sigma_{X_i}\right]^2}, \tag{2.5}$$

where $\frac{\partial Y}{\partial X_i}$ is the sensitivity coefficient, which is the partial derivative operator for the expectations of X_i, and σ_{X_i} is the standard deviation of the ith input variable X_i. The sensitivity coefficient is the result of a local linear approximation.

The PDF of Y is obtained by fitting a normal distribution to the expectation of Y as the mean and σ_Y as the standard deviation. This PDF can then be used to infer the uncertainty of Y given the known uncertainties of $\mathbf{X}$. This approach only requires the mean and standard deviation of the input quantities. It is thus the easiest and the most efficient method for uncertainty evaluation. It was consequently proposed as the standard for uncertainty evaluation by the *International Organization for Standardization* (ISO) [12].

Equation (2.5) is an analytical expression that provides the explicit relationship between the standard deviation of the output variable and the uncertainties of various input quantities. Its simplicity and analytic nature make it especially useful in design improvement and sensitivity analysis. As a result, this method has seen widespread adoption in numerous measurement and instrumentation applications, especially in cases in which the outputs of interest have low variance.

Despite its popularity, the use of linearization has an inherent deficiency in that it can be inaccurate when the variances are non-negligible with respect to the model's

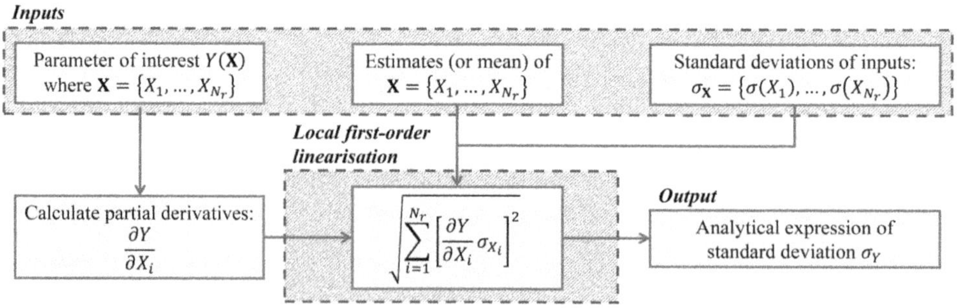

Figure 2.1. Framework of the first-order linearisation method used to evaluate the standard uncertainty of Y.

nonlinearity [13, 14]. In such cases, the GUM advocates that the higher-order Taylor series terms must be included to find the standard deviation of Y.

Highly nonlinear functions are often encountered in the design of engineering systems nearing their reliability limits. Unfortunately, the GUM does not outline a framework for variance estimation via higher-order Taylor series terms. In any case, this method inherently forces a normality assumption that unfortunately ignores high-order statistics such as skewness (a measure of asymmetry) and kurtosis (a measure of 'tailedness' or 'peakedness') [15]. As a result, this method is unreliable when there is a significant deviation from the normality assumption [16–18] (figure 2.2).

By way of example, consider a case in which a manufacturer needs to flatten steel plates from the initial thickness of h_0 to a desired thickness of h_1 via a flat rolling process [19]. The maximum change in the material thickness, denoted by ΔH, is measured using a system model. Using the statistical properties of ΔH reported in [19], the estimated PDF of ΔH is generated using the normality assumption (represented by the dotted blue line in figure 2.3). Another PDF that considers higher-order statistics (skewness and kurtosis) is also generated, which is represented as the continuous red line in the same figure. If manufacturers were to use the normal approximation approach with 99% confidence, they would have concluded that the maximum reduction in the material thickness is at least 1.21 cm per pass. However, such an analysis would be grossly incorrect. Referring to the PDF obtained with higher-order statistics, it is clear that the thickness can be reduced by at least 3.23 cm per pass with the same confidence level.

Based on this analysis, the number of passes through the plate-rolling machine can be reduced by 2.66 times. This would be hugely advantageous to the manufacturers, as it could potentially more than double the manufacturing yield of their steel plates without sacrificing the quality of their end product or requiring additional investment in more plate-rolling machines.

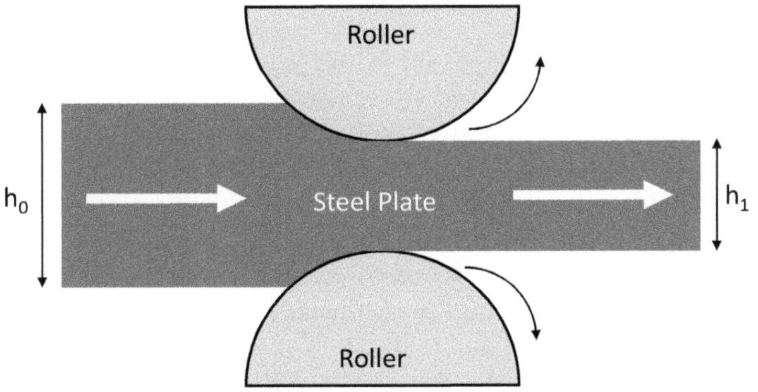

Figure 2.2. The plate rolling of steel, in which the flattening effect ΔH is measured from the system model, which includes the statistical properties of the material, the operational temperature, the roller size, the speed of rotation, and the steel plate thickness and movement [19].

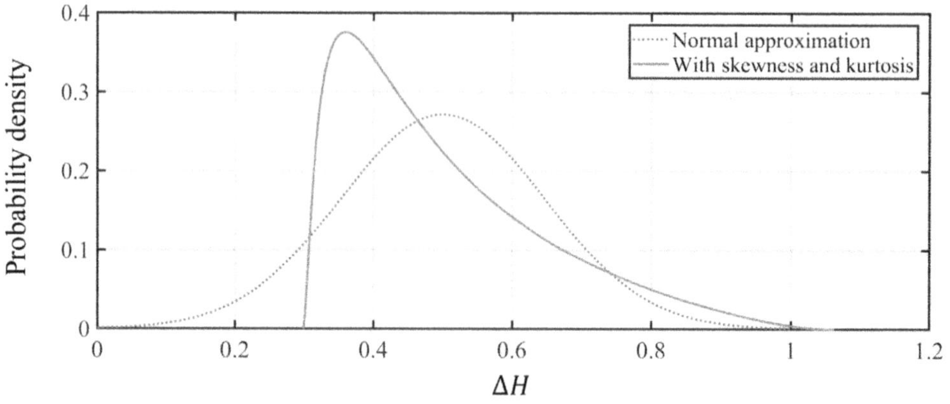

Figure 2.3. A comparison of PDF estimations for the maximum change in a material's thickness per pass ΔH. The dotted line is obtained using the normal approximation, whereas the continuous line shows the approximation obtained when higher-order statistics such as skewness and kurtosis are also considered.

The use of the normal approximation technique in such applications is therefore inadequate, and the decision-making arising from its use could have a substantial impact on cost, safety, and quality. In addition, if the standard deviation were wrongly approximated using local linearisation, the misrepresentation of the system would magnify this effect.

2.3 The Monte Carlo method

To address the limitations of first-order approximation, the *Monte Carlo* (MC) method has been utilised for uncertainty propagation [20]. The MC method simulates many independent realisations of the input quantities using the *a priori* probability distributions of the input quantities. The distributions are then propagated through $Y(\cdot)$ to find the output probability distribution from a histogram. Assuming that a sufficient number of trials are used, the MC method can reliably estimate the distribution of the output via simulation.

The MC simulation method is a very powerful approach and is immensely popular for uncertainty estimation due to the availability of cheap computational power and large memory resources. It enables the generation of any imaginable statistics about the output quantity, including the uncertainty, even in highly nonlinear systems. As a result, this method is valid for wider classes of uncertainty estimation problems compared to the first-order linearisation method with the normality assumption. Not only can the MC method model any output distribution, but its accuracy can also be increased reliably simply by increasing its sample size.

However, the MC method requires a sufficiently high number of realisations to achieve good probability convergence, and in turn, a reliable uncertainty estimation. More precisely, the variance in uncertainty estimation reduces in proportion to M^{-1}, where M is the sample size [21, 22]. Furthermore, unlike the previous approach, the MC uncertainty evaluation procedure is a purely numerical simulation process that does not produce an analytic expression of uncertainty; i.e. there is no explicit equation that describes the uncertainty of the output in terms of the inputs. This limits its use in the design optimisation process, which will be discussed further in section 3.2.2.

2.4 The use of a mathematical representation of the probability density function

The third method is based on analytical derivations that use the PDF. Various sources (e.g. [21, 23–26]) acknowledged the benefits of obtaining expressions for the output uncertainty in terms of the input uncertainties via this approach. However, within the context of design optimisation presented in this monograph, it is often excessively complex to derive the PDF of the output. Most engineers and designers would not be interested in spending precious development time deriving mathematical definitions for each possible model of their system.

Therefore, while this approach is more accurate compared to the first two methods outlined above, it remains a relatively unpopular one. To the best of the authors' knowledge, there has been no significant work to extend the analytical approach to uncertainty evaluation in general terms. This is perhaps because the majority of the reported cases are severely restrictive to a specific type of propagation equation and the particular distributions that can be used in those cases. Examples in the literature include the linear sum of normal or uniform distributions, the logarithmic transform of monotonous PDF [27], and the sum of the squares of normally distributed variables [20]. An approximate analytical treatment has also been reported for trapezoidal distributions [27, 28].

2.5 Uncertainty evaluation using moments

In cases in which the full PDF of an output Y cannot be directly expressed, it can instead be represented by its moments. This is especially convenient for standard uncertainty evaluation, which is defined in the GUM as the square root of the second-order central moment of a given measurement distribution [2]. According to this definition, only the first and second (raw) moments of Y are used to evaluate the standard uncertainty, given by equations (2.6) and (2.7).

The mean is given by:

$$\mu = \mathrm{E}[X]. \tag{2.6}$$

The standard deviation (or standard uncertainty) is given by:

$$\sigma = \sqrt{\mathrm{E}[X^2] - (\mathrm{E}[X])^2}. \tag{2.7}$$

Under the GUM framework, extended uncertainty is an extension of standard uncertainty to meet commercial, industrial, and regulatory requirements [2]. It can be obtained using a large fraction of the distribution of values that can reasonably be attributed to the measurand. Out of practical consideration, the concept of a coverage factor is introduced, in which the expanded uncertainty can be determined by multiplying the standard uncertainty by a corresponding coverage factor.

While the GUM suggests that the coverage factor derived from the normal distribution is sufficient under most circumstances, this assumption of sufficiency is rather subjective, as shown in [16–18].

In particular, the work presented in [18] uses Monte Carlo simulation on selected realistic measurement scenarios to demonstrate the bias or underestimation of the expanded uncertainty under this normality assumption. Furthermore, Fotowicz [17] shows that the quality of the expanded uncertainty estimation is unsatisfactory even when the coverage factors derived from simple distributions such as normal, trapezoidal, and uniform are used.

The assumption of a normal (or close to normal) distribution can be challenged when the underlying PDF has a high skew, a long tail, or both. In these cases, therefore, it is insufficient to use just the first two moments to characterise the PDF. Instead, the first four moments should be used [29], so that the skewness

$$\Omega = \frac{\mathrm{E}[X^3] - 3\mu\mathrm{E}[X^2] + 2\mu^3}{\sigma^3} \tag{2.8}$$

and kurtosis

$$\Psi = \frac{\mathrm{E}[X^4] - 4\mu\mathrm{E}[X^3] + 6\mu^2\mathrm{E}[X^2] - 3\mu^4}{\sigma^4} \tag{2.9}$$

are used in addition to the mean and the standard deviation shown in equations (2.6) and (2.7).

These higher-order moments $\mathrm{E}[X^n]$ are fed into various algorithmic estimations of parametric distributions, such as the normal [2], Cornish–Fisher expansion [30], extended generalised lambda [31], Tukey's *gh* [32], Pearson [33], Johnson [34], and maximum entropy [35] distributions. The confidence intervals can also be non-parametrically quantified using these moments. The use of parametric distribution fitting methods is a rather standard practice in applied statistics and engineering [36].

Although different distribution fitting techniques are available, comparing their goodness of fit (or fitting accuracy) is difficult without a common benchmarking platform. Section 5.2 introduces a set of benchmarking test distributions that can be used to assess the respective performances of these methods for reliability estimation in the probabilistic design optimisation procedure. The following subsections introduce some of the most well-known parametric fitting methods.

2.5.1 The normal distribution

As mentioned in the previous subsection, it is assumed in this method that the *normal (Gaussian)* distribution is a reasonably good fit to the data, and thus only the first two moments have to be included in the equation:

$$F_X(x) = \frac{1}{2}\left(1 + \mathrm{erf}\left(\frac{x - \mu_X}{\sigma_X\sqrt{2}}\right)\right), \tag{2.10}$$

where μ_X and σ_X are the mean and the standard deviation of input X. This is the method advocated by the GUM. It is also a theoretical basis that underpins the concept of a coverage factor. However, since the normal distribution is not skewed ($\Omega = 0$), it has been shown to perform poorly when the underlying distribution is skewed [29].

2.5.2 The Cornish–Fisher expansion

The *Cornish–Fisher* (CF) expansion originates from the classical mathematical probability theory [37] developed to describe a probability distribution that cannot be satisfactorily described by the normal distribution [38]. The CF fitting process starts with the normal distribution and then adjusts the distribution using additional information extracted from higher-order moments. This capability enables the approximation of distributions that do not significantly deviate from the normal distribution.

It should be noted that the CF approximation is not limited to four moments, even though that is the most common situation. The authors of this monograph examine the use of higher moments in the goodness-of-fit evaluation. In theory, if a moment can be computed from the parameters of a specific distribution, the maximum number of moments that can be used is restricted by the floating-point precision of the computer. However, a CF approximation implemented in 2015 utilised moments higher than the fourth order, as described in [30], but it failed to demonstrate a significant improvement in performance, as noted in [29]. Therefore, the discussion of CF approximation in this monograph is limited to the use of four moments only.

2.5.3 The extended generalised lambda distribution

The *generalised lambda distribution* (GLD) is a four-parameter continuous distribution [39]. The percentile function of the GLD is parameterised by λ_1, λ_2, λ_3, and λ_4, as shown in equation (2.11), where λ_1 represents the location parameter, λ_2 represents the scale parameter, and λ_3 and λ_4 control the behaviour of the left and right 'tails', respectively, while everything in between is a smooth interpolation of these two extremes.

$$F_X{}^{-1}(x) = \lambda_1 + \frac{x^{\lambda_3} - (1 - x)^{\lambda_4}}{\lambda_2}. \tag{2.11}$$

Equation (2.11) is designed to fit a wide variety of curve shapes [40] while remaining simple enough to be manipulated directly to find the values of the λ_i parameters for $i = 1, \ldots, 4$. The GLD has been widely employed since its introduction, and various improvements have been made, such as efficient parameter estimation and its application to reliability estimation [31, 40–43].

Although the GLD is capable of describing many important distributions of interest, a study [31] has shown that the GLD is unable to provide a high goodness of fit for distributions with high skew and low kurtosis. It went on to introduce the *generalised beta distribution* (GBD) to the GLD, resulting in a more powerful tool known as the *extended generalised lambda distribution* (EGLD). The GBD is a compact support distribution defined by its four-parameter PDF as:

$$f_X(x) = \frac{(x - \lambda_1)^{\lambda_3}(\lambda_1 + \lambda_2 - x)^{\lambda_4}}{\beta(\lambda_3 + 1, \lambda_4 + 1)\lambda_2^{\lambda_3+\lambda_4+1}}. \tag{2.12}$$

The EGLD is simpler, as it has only two distribution classes (GLD and GBD), while its parameter determination is much simpler. The parameters λ_i of the EGLD for $i = 1, \ldots, 4$ can easily be obtained from the table provided in [31] for the corresponding third- and fourth-order moments of the distribution.

The study described in [44] proposed another version of *extended generalised lambda distributions* (or XGLD) represented by:

$$F_X^{-1}(x) = \alpha_0 + \alpha_1 x^{\lambda_1} + \alpha_2(x - c)^{\lambda_2} - \alpha_3(1 - x)^{\lambda_3}, \tag{2.13}$$

where the distribution parameters α_0, α_1, α_2, λ_1, λ_2, λ_3, and c in expression (2.13) can be obtained using the moment matching method. The ith raw moment of X in the XGLD is given by:

$$\mathrm{E}[X^i] = \int_0^1 \left(\alpha_0 + \sum_{j=1}^{2} \alpha_j(x - c_j)^{\lambda_i}\right)^i \mathrm{d}x \tag{2.14}$$

for $c_1 = 0$, $c_2 = c$, and $c_3 = 1$ [44].

The XGLD is a more accurate approximation than the GLD if the initial condition is chosen correctly. Its improved accuracy is mainly attributed to a larger degree of freedom inherent in the model.

However, unlike the GLD, the parameter space of the XGLD is not well characterised. The XGLD formulation requires the use of numerical optimisation to perform the mapping from the moments to the distribution parameters. Thus, it causes significant interaction between the parameters and could quite likely lead to multiple local minima, making the XGLD unsuitable for uncertainty estimation.

2.5.4 Tukey's *gh* distribution

Tukey's gh distribution [45] is largely used to model the probability of extremely rare events, i.e. those corresponding to the tails of distributions [46], as well as other types of statistical modelling [47].

In common with the CF fitting technique, Tukey's *gh* fitting method starts with the standard normal distribution. The fitting is done by transforming the realisations x in (2.10) to z in:

$$z = \frac{\exp(gx) - 1}{g}\exp\left(\frac{hx^2}{2}\right), \tag{2.15}$$

where g and h are constant values of a real type.

The parameter g controls the skewness of the distribution with respect to the direction and the magnitude, while h controls the 'tail weight' of the distribution.

Therefore, for symmetrical distributions, $g = 0$ and only h has to be determined. The ith uncorrected moment of X is $\mathrm{E}[X^i]$, which can be given by [48]:

$$\mathrm{E}[X^i] = \left(g^i\sqrt{1 - ih}\right)^{-1}\sum_{j=0}^{i}(-1)^j\binom{i}{j}\exp\left(\frac{(i - j)^2 g^2}{2(1 - ih)}\right). \tag{2.16}$$

The skewness and kurtosis of X can be obtained by substitution into equations (2.8) and (2.9). In so doing, the values of g and h can be found by simultaneously solving for Ω and K with specific values of skewness and kurtosis [48]. Due to the high nonlinearity of the simultaneous equations, g and h have to be found using root-finding functions.

The downside of using Tukey's gh distribution is that it only accounts for $h > 0$. In other words, it can only transform the standard normal distribution into a heavier tail distribution (larger kurtosis), but not into a distribution with lower kurtosis. As a result, the performance of this method is expected to be similar to that of the normal approximation method for low kurtosis cases. Furthermore, the use of root-finding methods to solve for g and h imposes a high computational cost. Hence, to speed up the search for optimal g and h, a lookup table is provided in appendix A.

2.5.5 The Pearson distribution

Pearson was the first to introduce the use of skewness and kurtosis to characterise probability distributions in 1895 [49]. He then improved their definitions [50], and [51]. The *Pearson PDF* is defined by:

$$\frac{f_X{}'(x)}{f_X(x)} = \frac{x - a}{\lambda_0 + \lambda_1 x + \lambda_2\ x^2}, \tag{2.17}$$

where a, λ_0, λ_1, and λ_2 are the distribution parameters.

The system of Pearson distributions is the most well-known family of distributions. It covers many important unimodal distributions, including the normal distribution [36, 47]. It is a family of continuous probability distributions that are derived from a differential equation related to the normal density function.

The modern Pearson system contains seven distinct classes of distribution that can be determined based on the index value [47]:

$$\kappa = \frac{\lambda_1\ (\lambda_2 + 3)^2}{4(2\lambda_2 - 3\lambda_1 - 6)(4\lambda_2 - 3\lambda_1)}. \tag{2.18}$$

The PDFs of all seven types of Pearson distributions and their corresponding parameter criteria for classification are summarised in [36, 47]. The three main types of Pearson distribution are Type I, Type IV, and Type VI because they occupy areas in the (λ_1, λ_2) space, whereas the other types are represented by lines or points. Note that the normal distribution is a special case in the Pearson family of distributions where $\kappa = 0$, $\lambda_1 = 0$, and $\lambda_2 = 3$.

One of the attractive features of the Pearson distribution is that its parameters can be analytically determined from the first four moments. A combination of the parameters, in turn, selects a suitable distribution that best fits the observed moments.

2.5.6 The Johnson distribution

Just like Tukey's *gh* distribution, *Johnson distributions* [34, 52] use mathematical transformation functions to transform the realisations x into z, as follows:

$$z = \lambda_1 + \lambda_2 g\left(\frac{x - \lambda_3}{\lambda_4}\right), \tag{2.19}$$

where $g(\cdot)$ is a function of simple form, λ_1 and λ_2 are the shape parameters, λ_3 is the location parameter, and λ_4 is the scale parameter. Here, the transformation is performed based on one of the three functions shown in equations (2.20)–(2.22). They are lognormal

$$z = \lambda_1 + \lambda_2 \ln\left(\frac{x - \lambda_3}{\lambda_4}\right) \tag{2.20}$$

for $x \geqslant \lambda_3$, bounded

$$z = \lambda_1 + \lambda_2 \ln\left(\frac{x - \lambda_3}{\lambda_4 + \lambda_3 - x}\right) \tag{2.21}$$

for $\lambda_3 \leqslant x \leqslant \lambda_3 + \lambda_4$, and unbounded

$$z = \lambda_1 + \lambda_2 \sinh^{-1}\left(\frac{x - \lambda_3}{\lambda_4}\right) \tag{2.22}$$

for $-\infty < y < \infty$. For the sake of completeness, an identity transformation was added for the normal distribution; thus, the fourth transformation is:

$$z = \lambda_1 + \lambda_2\left(\frac{x - \lambda_3}{\lambda_4}\right) \tag{2.23}$$

for $-\infty < y < \infty$.

Here, the type of distribution that a set of moments belongs to is first determined using the skewness and kurtosis based on the Johnson curve [34]. Then, the first four statistical parameters of X are substituted into the general PDF:

$$f_X(x) = \frac{\lambda_2}{\sqrt{2\pi}} g'(x) \exp\left\{-\frac{1}{2}\left[\lambda_1 + \lambda_2 g\left(\frac{x - \lambda_3}{\lambda_4}\right)\right]\right\}. \tag{2.24}$$

As in the Pearson distribution, the first four moments determine the analytical PDF from the family of distributions. Therefore, the Johnson distribution can be regarded as a crossover between Tukey's *gh* method and Pearson method.

2.5.7 The maximum entropy method

For a random variable X whose realisation x takes all values over an interval of real numbers with a unique PDF $f_X(x)$, the Shannon entropy S [53] is defined as:

$$S = -\int f_X(x)\ln\left[f_X(x)\right]\mathrm{d}x. \tag{2.25}$$

The maximum entropy algorithm [35] uses monomials (i.e. x^i) as the basis functions in expressions (2.26)–(2.30). In the moment-based maximum entropy method, the information entropy S is maximised subject to (2.25) for $i = 0, \dots, N_m$ [54]. Using the method of Lagrange multipliers [55], the optimisation problem with $N_m + 1$ constraints (considering the zeroth moment) is then reduced to the optimisation of the unconstrained function:

$$\mathcal{L}(\boldsymbol{\varphi}) = \int \exp\left(\sum_{i=0}^{N_m}\varphi_i x^i\right)\mathrm{d}x - \sum_{i=0}^{N_m}\varphi_i x^i, \tag{2.26}$$

where $\boldsymbol{\varphi} = \left\{\varphi_0, \lambda\varphi_1, \dots, \varphi_{N_m}\right\}$ is the Lagrange multiplier.

Expression (2.26) has a closed-form solution for $f_X(x)$:

$$f_X(x) = \exp\left(\sum_{i=0}^{N_m}\varphi_i x^i\right). \tag{2.27}$$

All the optimal maximum entropy distributions are achieved when $\partial\mathcal{L}/\partial\boldsymbol{\varphi} \cong 0$, which automatically satisfies the moment constraints in (2.25) and takes the general form $f_X(x)$ in (2.27). Note that since $x^0 = 1$ (for $i = 0$) in (2.30), φ_0 can be found using the explicit function:

$$\varphi_0 = -\ln\int_{\mathbb{R}} \exp\left(\sum_{i=1}^{N_m}\varphi_i x^i\right)\mathrm{d}x. \tag{2.28}$$

The gradient $(\nabla\mathcal{L})_i$ and the elements H_{ij} of the Hessian matrix $\mathbf{H}$ of the Lagrangian function in (2.26) are given, respectively, by:

$$(\nabla\mathcal{L})_i = \frac{\delta\mathcal{L}}{\delta\varphi_i} = \int_{\mathbb{R}} x^i f_X(x)\mathrm{d}x - x^i, \tag{2.29}$$

$$H_{ij} = \int_{\mathbb{R}} x^{i+j} f_X(x)\mathrm{d}x \tag{2.30}$$

According to the principle of maximum entropy [56], all likelihoods are to be considered based on the information available in the data, which in this case is the set of truncated moments associated with the distribution of interest. The maximum entropy distribution is the one with the largest overall uncertainty out of all possible distributions with the same moment sequence. Theoretically, the higher the number of moments used in the maximum entropy method, the closer the estimated PDF is to the actual distribution, since more information from the moments diminishes the uncertainty in the maximum entropy distribution.

However, as the number of moments increases (especially when $N_m > 5$), the procedure becomes highly sensitive to any numerical imbalance in the moments, an ill-conditioned gradient or Hessian matrix, and insufficient arithmetic precision [57, 58]. To address these shortcomings, advanced and numerically stable algorithms were developed using different basis functions, such as the shifted Chebyshev polynomials [57] and the Fup functions [58]. The work described in [59] further extends the improvements to multidimensional problems using *generalised orthogonal polynomials* (GOPoly).

A final shortcoming of these algorithms is that they assume that the range of the distribution is known. Convergence is not guaranteed in an automatic application of the procedures if the limit of integration is not (at least approximately) known beforehand [57–59]. This limitation will be addressed in chapter 5.

2.6 Summary

This chapter on uncertainty propagation has covered three methods for estimating the measurand and evaluating uncertainty in nonlinear models: Method 1, Method 2, and Method 2S. These methods focus on assigning probability distributions to quantities in the measurement model. This chapter explored the common methods used in uncertainty evaluation: first-order linearisation with normal approximation, Monte Carlo simulation, analytical methods using mathematical PDF representations, and uncertainty evaluation using moments. It provided a detailed examination of distribution fitting methods such as the normal, Cornish–Fisher, extended generalised lambda, Tukey's *gh*, Pearson, Johnson, and maximum entropy methods.

References

[1] ISO/IEC Guide 98–6:2021 2021 *Uncertainty of Measurement—Part 6: Developing and Using Measurement Models* (International Organisation for Standardization)

[2] Guides in Metrology Committee: JCGM 2008 *Evaluation of measurement data—guide to the expression of uncertainty in measurement* (Sèvres: BIPM) https://www.bipm.org/en/committees/jc/jcgm/publications JCGM 100:2008

[3] Xiu D and Karniadakis G E 2002 The Wiener–Askey polynomial chaos for stochastic differential equations *SIAM J. Sci. Comput.* **24** 619–44

[4] Xiu D and Karniadakis G E 2002 Modeling uncertainty in steady state diffusion problems via generalized polynomial chaos *Comput. Meth. Appl. Mech. Eng.* **191** 4927

[5] Ghanem R and Spanos P D 1990 Polynomial chaos in stochastic finite elements *J. Appl. Mech.* **57** 197–202

[6] Macii D, Mari L and Petri D 2008 Measurand value estimation in nonlinear models: a comparison between two averaging methods *2008 IEEE Int. Workshop on Advanced Methods for Uncertainty Estimation in Measurement* (Piscataway, NJ: IEEE) 22–7

[7] Macii D, Mari L and Petri D 2009 Comparison of measured quantity value estimators in nonlinear models *IEEE Trans. Instrum. Meas.* **59** 238–46

[8] Guides in Metrology Committee: JCGM 2012 *International Vocabulary of Metrology - Basic and general concepts and associated terms (2012)* 3 (Sèvres: BIPM) https://www.bipm.org/en/committees/jc/jcgm/publications JCGM 200:2012. This 3rd edition is also published as an ISO guide.
[9] Bich W, Callegaro L and Pennecchi F 2006 Non-linear models and best estimates in the GUM *Metrologia* **43** S196
[10] Wang C-M and Iyer H K 2011 On non-linear estimation of a measurand *Metrologia* **49** 20
[11] Kuang Y C, Rajan A, Ooi M P-L and Demidenko S N 2017 Measured quantity value estimator for multiplicative nonlinear measurement models *IEEE Trans. Instrum. Meas.* **66** 715–22
[12] ISO/IEC Guide 98-1:2009 2009 *Uncertainty of measurement—Part 1: Introduction to the expression of uncertainty in measurement* (International Organisation for Standardization)
[13] Pendrill L R 2014 Using measurement uncertainty in decision-making and conformity assessment *Metrologia* **51** S206
[14] Ribeiro A S, Sousa J A e, Costa C O, Castro M P and Cox M G 2008 Uncertainty evaluation and validation of a comparison methodology to perform in-house calibration of platinum resistance thermometers using a Monte Carlo method *Int. J. Thermophys.* **29** 902–14
[15] Ayyub B M 1997 *Uncertainty Modeling and Analysis in Civil Engineering* (Boca Raton, FL: CRC Press)
[16] Possolo A 2014 Statistical models and computation to evaluate measurement uncertainty *Metrologia* **51** S228
[17] Fotowicz P 2004 A method of approximation of the coverage factor in calibration *Measurement* **35** 251–6
[18] Vilbaste M, Slavin G, Saks O, Pihl V and Leito I 2010 Can coverage factor 2 be interpreted as an equivalent to 95\% coverage level in uncertainty estimation? Two case studies *Measurement* **43** 392–9
[19] Anderson T V and Mattson C A 2012 Propagating skewness and kurtosis through engineering models for low-cost, meaningful, nondeterministic design *J. Mech. Des.* **134** 100911
[20] Guides in Metrology Committee: JCGM 2008 *Evaluation of measurement data—supplement 1 to the guide to the expression of uncertainty in measurement—propagation of distributions using a Monte Carlo method* (Sèvres: BIPM) https://www.bipm.org/en/committees/jc/jcgm/publications JCGM 101:2008
[21] Cox M G and Siebert B R L 2006 The use of a Monte Carlo method for evaluating uncertainty and expanded uncertainty *Metrologia* **43** S178
[22] Harris P M and Cox M G 2014 On a Monte Carlo method for measurement uncertainty evaluation and its implementation *Metrologia* **51** S176
[23] Bertrand J, Bertrand P and Ovarlez J-P 2000 The Mellin transform *The Transforms And Applications Handbook* (Boca Raton, FL/Piscataway, NJ: CRC Press/IEEE Press)
[24] Cox M G and Harris P M 2014 Validating the applicability of the GUM procedure *Metrologia* **51** S167
[25] Lomnicki Z A 1967 On the distribution of products of random variables *J. R. Stat. Soc. Ser. B: Methodol.* **29** 513–24
[26] Springer M D and Thompson W E 1966 The distribution of products of independent random variables *SIAM J. Appl. Math.* **14** 511–26

[27] Cox M G and Harris P M 2006 *Software support for metrology best practice guide no. 6: uncertainty and statistical modelling* (Teddington: NPL) https://eprintspublications.npl.co.uk/3879/
[28] Dietrich C 1991 *Uncertainty, Calibration and Probability: The Statistics of Scientific and Industrial Measurement* (Boca Raton, FL: CRC Press)
[29] Kuang Y C, Ooi M P-L, Rajan A and Demidenko S 2015 Performance comparison between expanded uncertainty evaluation algorithms *2015 IEEE International Instrumentation and Measurement Technology Conference (I2MTC) Proceedings* (Piscataway, NJ: IEEE) 1729–34
[30] Lee Y-S and Lin T-K 1992 Algorithm AS 269: high order Cornish–Fisher expansion *J. R. Stat. Soc. Ser. C: Appl. Stat.* **41** 233–40
[31] Karian Z A and Dudewicz E J 2000 *Fitting Statistical Distributions: The Generalized Lambda Distribution and Generalized Bootstrap Methods* (New York: Chapman and Hall/CRC)
[32] Jorge M and Boris I 1984 Some properties of the Tukey *g* and *h* family of distributions *Commun. Stat. – Theory Meth.* **13** 353–69
[33] Nagahara Y 2004 A method of simulating multivariate nonnormal distributions by the Pearson distribution system and estimation *Comput. Stat. Data Anal.* **47** 1–29
[34] Johnson N L, Kotz S and Balakrishnan N 1994 *Continuous Univariate Distributions* 2 (Hoboken, NJ: Wiley)
[35] Mead L R and Papanicolaou N 1984 Maximum entropy in the problem of moments *J. Math. Phys.* **25** 2404–17
[36] Xi Z, Hu C and Youn B D 2012 A comparative study of probability estimation methods for reliability analysis *Struct. Multidiscip. Optim.* **45** 33–52
[37] Cornish E A and Fisher R A 1938 Moments and cumulants in the specification of distributions *Rev. Int. Stat. Inst.* **5** 307–20
[38] Fisher S R A and Cornish E A 1960 The percentile points of distributions having known cumulants *Technometrics* **2** 209–25
[39] Ramberg J S and Schmeiser B W 1974 An approximate method for generating asymmetric random variables *Commun. ACM* **17** 78–82
[40] Su S 2007 Numerical maximum log likelihood estimation for generalized lambda distributions *Comput. Stat. Data Anal.* **51** 3983–98
[41] Asquith W H 2007 L-moments and TL-moments of the generalized lambda distribution *Comput. Stat. Data Anal.* **51** 4484–96
[42] Fournier B, Rupin N, Bigerelle M, Najjar D, Iost A and Wilcox R 2007 Estimating the parameters of a generalized lambda distribution *Comput. Stat. Data Anal.* **51** 2813–35
[43] Karvanen J and Nuutinen A 2008 Characterizing the generalized lambda distribution by L-moments *Comput. Stat. Data Anal.* **52** 1971–83
[44] Ahmadabadi M N, Farjami Y and Moghadam M B 2012 Approximating distributions by extended generalized lambda distribution (XGLD) *Commun. Stat. – Simul. Comput.* **41** 1–23
[45] Tukey J W 1977 Modern techniques in data analysis *Proc. of the NSF-Sponsored Regional Research Conf.* 7 *(Southeastern Massachusetts University, North Dartmouth, MA, 1977)*
[46] Headrick T C, Kowalchuk R K and Sheng Y 2008 Parametric probability densities and distribution functions for Tukey g-and-h transformations and their use for fitting data *Appl. Math. Sci.* **2** 449–62 http://www.m-hikari.com/ams/ams-password-2008/ams-password9-12-2008/headrickAMS9-12-2008.pdf

[47] Sim C H and Lim M H 2008 Evaluating expanded uncertainty in measurement with a fitted distribution *Metrologia* **45** 178
[48] Kowalchuk R K and Headrick T C 2010 Simulating multivariate *g*-and-*h* distributions *Br. J. Math. Stat. Psychol.* **63** 63–74
[49] Pearson K 1895 Contributions to the mathematical theory of evolution. II. Skew variation in homogeneous material *Philos. Trans. R. Soc. London* **186** 343–424
[50] Pearson K 1901 XI. Mathematical contributions to the theory of evolution.—X. Supplement to a memoir on skew variation *Philos. Trans. R. Soc. London Ser.* A **197** 443–59
[51] Pearson K 1916 IX. Mathematical contributions to the theory of evolution.—XIX. Second supplement to a memoir on skew variation *Philos. Trans. R. Soc. London Ser.* A **216** 429–57
[52] Johnson N L 1949 Systems of frequency curves generated by methods of translation *Biometrika* **36** 149–76
[53] Shannon C E 2001 A mathematical theory of communication *ACM SIGMOBILE Mob. Comput. Commun. Rev.* **5** 3–55
[54] Rajan A, Kuang Y C, Ooi M P-L, Demidenko S N and Carstens H 2018 Moment-constrained maximum entropy method for expanded uncertainty evaluation *IEEE Access* **6** 4072–82
[55] Bertsekas D P 1999 *Nonlinear Programming* (Amsterdam: Athena Scientific)
[56] Jaynes E T 1957 Information theory and statistical mechanics *Phys. Rev.* **106** 620
[57] Bandyopadhyay K, Bhattacharya A K, Biswas P and Drabold D A 2005 Maximum entropy and the problem of moments: a stable algorithm *Phys. Rev.* E **71** 057701
[58] Gotovac H and Gotovac B 2009 Maximum entropy algorithm with inexact upper entropy bound based on Fup basis functions with compact support *J. Comput. Phys.* **228** 9079–91
[59] Abramov R V *et al* 2010 The multidimensional maximum entropy moment problem: a review of numerical methods *Commun. Math. Sci.* **8** 377–92

Chapter 3

Probabilistic design optimisation

Design optimisation is an iterative process used to determine the design parameters of a system that maximise the performance of the design while meeting a set of prescribed physical constraints. Traditional methods of design optimisation were generally deterministic and did not consider the uncertainties in the design parameters [1]. Figure 3.1(a) is an illustration of a generic optimisation framework. An optimisation algorithm is implemented to iteratively update the design point in Step 4.

In today's modern competitive technological market, there is a push to produce more economical and sustainable designs that meet increasingly stringent safety and quality requirements. For example, for the sensors produced today, the parameter variations due to manufacturing uncertainties must be accounted for, otherwise their manufacturers risk producing large amounts of non-compliant products, leading to either a loss of sales or the risk of incurring additional costs due to non-compliance [2]. In more mission-critical applications such as structural systems, design engineers must simultaneously adhere to engineering building codes while also ensuring the use of minimal resources to meet financial and sustainability goals [3].

Reliability and/or robustness analysis is increasingly employed within design optimisation procedures to ensure that uncertainties are considered within the design loop—especially so when health and safety factors (e.g. structural buildings and bridges [4]) or large-scale component manufacture (e.g. sensors and actuators [5]) are involved. This comes in the form of additional steps in the design optimisation framework that performs probabilistic analyses of reliability and robustness (figure 3.1(b)). Here, all design parameter uncertainties and their propagation through the system are considered by computing the probabilistic constraint of the design point in that iteration. In so doing, the resultant design should be the most economical design that maximises performance and meets the physical constraints at a specified level of confidence (reliability) and its susceptibility to uncertainty

doi:10.1088/978-0-7503-4931-4ch3

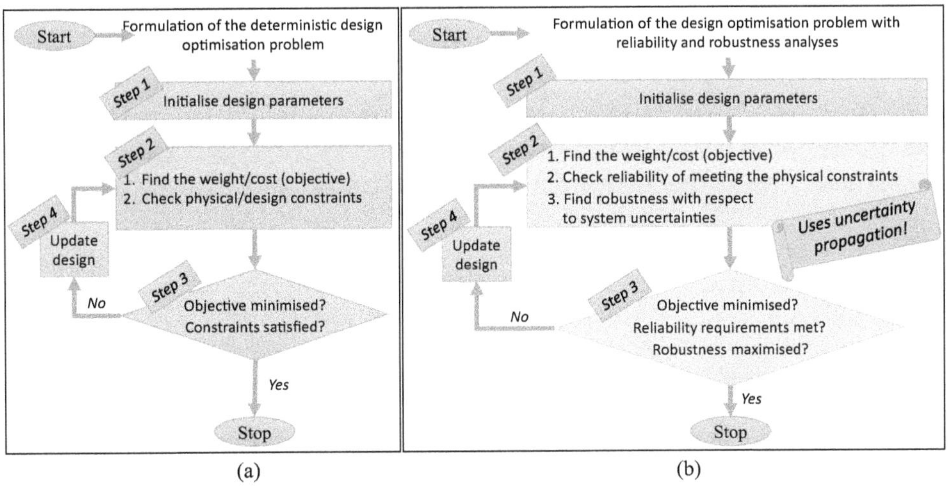

Figure 3.1. Comparison of (a) deterministic and (b) probabilistic design optimisation frameworks, highlighting the role of uncertainty propagation.

(robustness). This allows the system designer or engineer to ensure that the resultant design is compliant with the specified reliability and safety requirements.

Figure 3.1 highlights the key differences between the deterministic and probabilistic design optimisation strategies. What is perhaps not as obvious from figure 3.1 is that the optimisation algorithm sits within Step 4. Different optimisation algorithms can be deployed with different levels of convergence efficiency. This topic itself is still an active research field and falls outside the scope of this book. Instead, this book uses state-of-the-art general-purpose optimisation algorithms such as sequential quadratic programming [6] and genetic algorithms [7]. It focuses solely on the formulation of the design optimisation framework, how uncertainty propagation can be incorporated into it more effectively, how the parameters are computed (Step 2), and how the optimum design is determined (Step 3).

Design optimisation typically requires the specification of two classes of parameter, which are the objective function(s) and the physical constraint functions. In probabilistic design optimisation, the constraints of a system's performance are represented using performance functions. In general, the desired reliability is predetermined within the optimisation formulation based on safety, regulatory, or compliance standards. The robustness of a chosen constraint may either be omitted or optimised together with the design objective. The reliability limit assigned to each performance function is known as the reliability constraint [8]. The biggest challenge in probabilistic design optimisation is to evaluate these probabilistic constraints in a way that is simultaneously efficient and accurate. In general, the exact computation of these reliability constraints is too complex to be performed for multivariate nonlinear systems. Instead, numerical approximations are employed, such as computing only the first- and second-order derivatives of the performance functions, thus sacrificing accuracy for efficiency.

The differences between reliability and robustness analyses are shown in figure 3.2. In this figure, y denotes the realisation of output Y and there is a constraint function

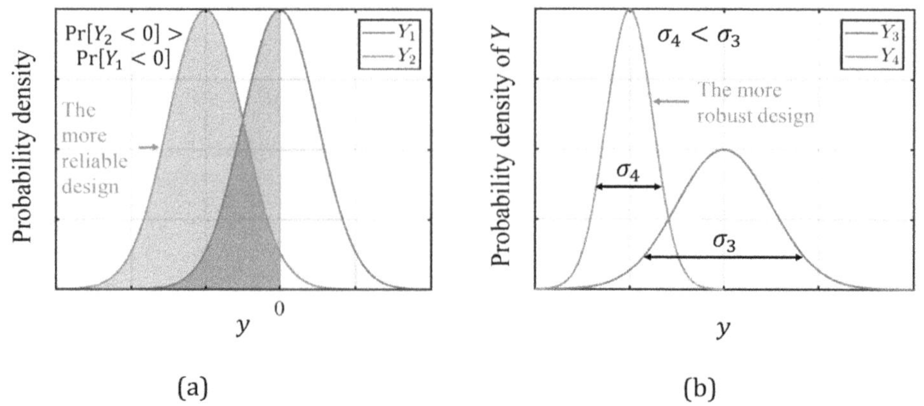

Figure 3.2. A comparison of the PDFs of output Y for (a) reliable and (b) robust designs. Pr[·] denotes the probability operator, σ denotes the standard deviation, and y denotes the realisation of Y.

of $Y \leqslant 0$. At every iteration within the optimisation procedure, a new design point is obtained. These are shown as Y_1, Y_2, Y_3, and Y_4 (each is a distinct design variation of Y). For reliability analysis, the probability of $Y \leqslant 0$ must be verified at every design point during the iterative optimisation procedure. Figure 3.2(a) shows a comparison of the *probability density functions* (PDFs) of two design iterations Y_1 and Y_2. Observing that the peak of the PDF of Y_2 is further to the left of the $Y = 0$ constraint than that of the PDF of Y_1, it is clear that the design parameters corresponding to Y_2 have a higher probability of meeting the $Y \leqslant 0$ constraint function compared to those of Y_1. In short, one can surmise that the parameters of Y_2 produce a more reliable design.

The robustness is determined based on the spread of the PDF. A smaller spread of the distribution demonstrates its lower susceptibility to uncertainties in the design parameters. This means that the corresponding design would have a more consistent output. Figure 3.2(b) shows the PDFs for two design iterations Y_3 and Y_4. It can be seen that Y_3 has a larger spread than that of Y_4; therefore, it can be concluded that the design parameters corresponding to Y_4 result in a more robust design compared to those of Y_3.

The constraint functions themselves can be difficult to establish. This challenge originates from the complexity of the system and is especially prevalent in numerous engineering applications where the output of interest Y is a conceptual construct that cannot be directly measured. This means that its uncertainty cannot be determined via measurement. Instead, a model-based determination $Y(\mathbf{X})$ is used. This allows Y to be determined using the estimates of a set of N_r random variables $\mathbf{X} = \{X_1, \ldots, X_{N_r}\}$ through a known (or assumed) mathematical function $Y(\cdot)$. In most realistic scenarios, the function $Y(\cdot)$ may be defined implicitly. This will be elaborated with real-world examples in chapter 6.

The statistical properties of $\mathbf{X}$ can be established in several ways. They can be determined via measurements, obtained from the manufacturer's specification documents, derived from a physical understanding, or even deduced from *a priori* assumptions. The known uncertainties of **X** can be then propagated through function $Y(\cdot)$ to find the PDF or *cumulative distribution function* (CDF) of the constraint functions; consequently, the uncertainties in Y can be evaluated. The uncertainties can then be used to perform reliability and robustness analyses in Step 2 of the probabilistic design optimisation shown in figure 3.1(b).

Obtaining the PDFs or CDFs of the constraint functions is therefore crucial for evaluating their uncertainties and in turn performing reliability and robustness analyses in probabilistic optimisation. However, as discussed in chapter 2, it is impractical if not impossible to mathematically derive all of the probability distributions for each possible system output analytically. Therefore, an effective method for approximating the probability distributions of Y is needed. The *propagation of uncertainty* [9], i.e. the process of inferring output uncertainties (typically from the PDF or CDF of Y) given known input uncertainties was extensively discussed in the previous chapter.

In a deterministic design optimisation problem in which the inherent uncertainties are not considered, the design problem can be expressed mathematically as:

$$\begin{aligned} &\text{minimise: } C(\mathbf{d}), \\ &\text{subject to: } G_i(\mathbf{d}) \leqslant 0 \ \text{ for } i = 1,\ldots,N_c, \\ &\text{where: } \quad \mathbf{d}^L \leqslant \mathbf{d} \leqslant \mathbf{d}^U, \end{aligned} \tag{3.1}$$

where $C(\cdot)$ is the objective function to be minimised, $\mathbf{d} = \{d_1,\ldots,d_{N_d}\}$ is a vector of the N_d design variables with a lower bound $\mathbf{d}^L$ and an upper bound $\mathbf{d}^U$, $G_i(\cdot)$ is the ith constraint function (which represents the physical performance criterion to be satisfied), and N_c is the number of constraint functions. Some examples of $C(\cdot)$ include the weight of a structure, the cost of the product, the performance variance, etc.

The probabilistic design optimisation paradigms that are commonly used in engineering design are: (1) *robust design optimisation* (RDO); (2) *reliability-based design optimisation* (RBDO); and (3) *reliability-based robust design optimisation* (RBRDO), which combines the first two frameworks. The following subsections present brief descriptions of each of these formulations.

3.1 Robust design optimisation

A product or system designed by a robust design optimisation framework is expected to be insensitive to external noise or tolerances [10]. Since all materials and processes come with their associated uncertainty (such as modulus, thickness, density, noise, and vibration), the robust design optimisation framework includes sensitivity analysis within the optimisation procedure. It does so by minimising the sensitivity of the design objectives to such variations within each design iteration. Figure 3.3 shows the different types of performance variations.

The moment-based method [12–15] of assessing system robustness is more likely to be employed than other techniques such as the percentile difference-based method [16] or the hybrid quality loss functions-based method [17] due to its superior

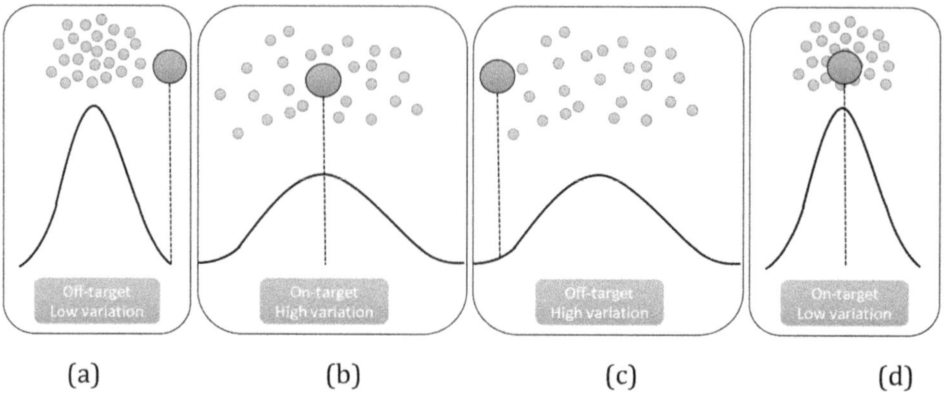

Figure 3.3. A depiction of different types of performance variation within a system [11].

accuracy and computational efficiency [13]. Moment-based robust design optimisation is essentially a modified version of the notation in (3.1), given by:

$$\begin{aligned} &\text{minimise:}\ C\left(\mu_K,\ \sigma_K{}^2\right),\\ &\text{subject to:}\ G_i(\mathbf{d}) \leqslant 0\ \text{ for } i = 1,\ldots,N_c,\\ &\text{where:}\quad \mathbf{d}^L \leqslant \mathbf{d} \leqslant \mathbf{d}^U, \end{aligned} \tag{3.2}$$

where the objective function $C(\cdot)$ from equation (3.1) is modified to express the performance function $K(\mathbf{d}, \mathbf{X})$ in terms of its mean μ_K and variance $\sigma_K{}^2$.

The performance function quantifies the capability of a system under design with respect to the design variables **d** and random variables **X**. Often, one of the constraint functions is chosen to be the performance function. For example, in the assessment of vehicle side impact crashworthiness [18], the post-impact lower rib area deflection of the dummy is the performance function as well as the constraint function. In the case of this example, a safer design can be obtained by minimising the variance of the rib area deflection in addition to meeting the required safety standards.

The mean and variance of the performance function can be calculated from the first two moments. The ith-order moments of $K(\mathbf{d}, \mathbf{X})$, denoted by $\mathrm{E}[K^i]$, can be calculated using the multidimensional integral:

$$\mathrm{E}[K^i] = \int \cdots \int K^i(\mathbf{d}, \mathbf{X}) f_{\mathbf{X}}(\mathbf{x}) \mathrm{d}\mathbf{x}, \tag{3.3}$$

where $\mathbf{x} = \{x_1,\ldots,x_{N_r}\}$ denotes realisations of the random variables **X** and $f_{\mathbf{X}}(\mathbf{x})$ is the joint PDF of **X** expressed in terms of **x**. Here, the vector **d** may be either independent of, or linked to the random variable vector **X**. In most design problems, $\mathbf{d} = \mathrm{E}[\mathbf{X}]$, where $\mathrm{E}[\mathbf{X}]$ is the expectation of the random variable **X**.

There are two mainstream computational methods for decomposing the performance function $K(\cdot)$ in equation (3.3), which are the *univariate dimension reduction method* and the *performance moment integration method*.

3.1.1 The univariate dimension reduction method

A performance function $K(\mathbf{X})$ with N_r-dimensional input random variables $\mathbf{X}$ can be additively decomposed into one-dimensional functions using the univariate dimension reduction method shown in (3.4) [12]:

$$K(\mathbf{X}) \cong \sum_{i=1}^{n_{rv}} h\left(\mu_{X_1}, \ldots, \mu_{X_{i-1}}, X_i, \mu_{X_{i-1}}, \ldots, \mu_{X_{N_r}}\right) - (N_r - 1)h(\mu_{X_1}, \ldots, \mu_{X_{N_r}}), \quad (3.4)$$

where N_r is the number of random variables, and μ_{X_i} denotes the mean of random variable X_i for arbitrary i.

Decomposition (3.4) is valid for all functions of $K(\cdot)$ with a convergent Taylor series expansion centered at origin (i.e, 0). [15]. As a result of this decomposition, the multidimensional integration in (3.4) is reduced to N_r one-dimensional integrations, which are then numerically calculated using the *moment-based integration rule* (MBIR). Paper [15] outlines the MBIR procedure used to calculate the required weights w_i^j and quadrature points x_i^j depending on the distribution type and the number of quadrature points N_q. In short, the general expression for evaluating the mean and the variance of the performance function $K(\cdot)$ using the univariate DRM method can be written as (3.5) and (3.6), respectively:

$$\mu_K \cong \sum_{j=1}^{N_q} \sum_{i=1}^{N_r} w_i^j K\left(\mu_1, \ldots, \mu_{X_{i-1}}, x_i^j, \mu_{X_{i-1}}, \ldots, \mu_{X_{N_r}}\right) - (N_r - 1)K(\mu_1, \ldots, \mu_{X_{N_r}}) \quad (3.5)$$

$$\begin{aligned}\sigma_K{}^2 \cong & \sum_{j=1}^{N_q} \sum_{i=1}^{N_r} w_i^j K^2\left(\mu_1, \ldots, \mu_{X_{i-1}}, x_i^j, \mu_{X_{i-1}}, \ldots, \mu_{X_{N_r}}\right). \\ & - (N_r - 1)K^2(\mu_1, \ldots, \mu_{X_{N_r}}) - \mu_K{}^2\end{aligned} \quad (3.6)$$

3.1.2 The performance moment integration method

The performance moment integration method uses the Rosenblatt transformation [19] to rewrite the multidimensional integral of equation (3.3) as a one-dimensional integral [14]:

$$\mathrm{E}[K^i] = \int K^i(t; \boldsymbol{\mu}_{\mathbf{X}})\Phi(t)\mathrm{d}t, \quad (3.7)$$

where $\Phi(\cdot)$ is the normal cumulative distribution function and t is the distance from the origin to the *most probable point* (MPP) in the normalised u-space.

It then makes use of weights and quadrature points to evaluate the transformed integral in equation (3.7). Since t follows the standard normal distribution in the u-space, the weights and quadrature points for $N_q = 3$ are used to further reduce equation (3.7) to the expression shown in (3.8).

$$\begin{aligned} \mathrm{E}[K^i] &\cong \frac{1}{6}K^i(t_1; \boldsymbol{\mu}_X)\Big|_{t_1=-\sqrt{3}} + \frac{4}{6}K^i(t_2; \boldsymbol{\mu}_X)\Big|_{t_2=0} + \frac{1}{6}K^i(t_3; \boldsymbol{\mu}_X)\Big|_{t_3=\sqrt{3}} \\ &= \frac{1}{6}K^i\left(-\sqrt{3}; \boldsymbol{\mu}_X\right) + \frac{4}{6}K^i(0; \boldsymbol{\mu}_X) + \frac{1}{6}K^i\left(\sqrt{3}; \boldsymbol{\mu}_X\right) \end{aligned} \tag{3.8}$$

Using special algorithms such as the hybrid mean value [20] and conjugate gradient analysis [21] methods, each term in equation (3.8) can be approximated. Using equation (3.8), the mean $\mu_K = \mathrm{E}[K]$ and variance $\sigma_K{}^2 = \mathrm{E}[K^2] - (\mathrm{E}[K])^2$ of the performance function then can be calculated for robustness analysis.

3.2 Reliability-based design optimisation

Reliability-based design optimisation considers the design objectives for a given set of probabilistic constraints for the constraint functions $G(\cdot)$ at every iteration. The RBDO problem can be mathematically expressed as a modification of equation (3.1):

$$\begin{aligned} &\text{minimise: } C(\mathbf{d}), \\ &\text{subject to: } \Pr[G_i(\mathbf{d}, \mathbf{X}) \leqslant 0] > \Phi(\beta_i) \text{ for } i = 1,\ldots,N_c \\ &\text{where: } \quad \mathbf{d}^L \leqslant \mathbf{d} \leqslant \mathbf{d}^U, \end{aligned} \tag{3.9}$$

where $C(\cdot)$ is the objective function, $\Phi(\cdot)$ is the standard normal cumulative distribution function, $\Pr[\cdot]$ is the probability operator, and β_i is the ith target reliability index corresponding to the performance function $G_i(\cdot)$. The probability expression $\Pr[G_i(\mathbf{d}, \mathbf{X}) \leqslant 0]$ in (3.9) can be explicitly expressed as:

$$\Pr[G_i(\mathbf{d}, \mathbf{X}) \leqslant 0] = \int \ldots \int_{G_i(\mathbf{d},\mathbf{X})\leqslant 0} f_{\mathbf{X}}(\mathbf{x})\mathrm{d}\mathbf{x}, \tag{3.10}$$

where $f_{\mathbf{X}}(\mathbf{x})$ is the joint probability density function of $\mathbf{X}$. As discussed earlier in this chapter, numerical approximations are popularly used in reliability analysis because of their mathematical complexity. These include techniques such as:

- *Most probable point* (MPP)-based methods [4, 22, 23], which are considered state of the art because of their simplicity and efficiency;
- *Metamodels* (or surrogate models) [24–28] that approximate the input–output relationship using direct sampling methods such as the *MC* method [24] and *importance sampling* [25];
- *Moment-based* methods [29–33] calculate statistical moments up to the 4th order. This book will expand the moment-based method given in section 3.2.4 to describe a new RBDO technique which combines a high-order analytical moments method based on the Mellin transform, local response surface modelling using polynomial genetic programming, and a moment-constrained entropy method.

3.2.1 Most probable point-based methods

The MPP can be interpreted as the most likely point of failure conforming to a reliability index β, which in turn corresponds to the standard normal inverse cumulative distribution with a known probability. For example, for a probability of 0.9987 (or 99.87%), the value of β is 3. MPP-based approaches such as the *first- and second-order reliability methods* (FORM and SORM, respectively) [18, 19, 45, 46] bypass the uncertainty evaluation by applying a probabilistic transformation such as the Rosenblatt transformation [19] to transform the realisation of every non-normal random variable from an original x-space into a standardised normal distribution space known as the u-space. According to the transformation, for a point x, the equivalent point in u-space is found by solving:

$$F_{\mathbf{X}}(\mathbf{x}) = \Phi(\mathbf{u}), \tag{3.11}$$

where $F_{\mathbf{X}}(\cdot)$ is the CDF of $\mathbf{X}$ defined with respect to realisations $\mathbf{x}$ and $\mathbf{u} = \{u_1,\ldots,u_{N_r}\}$ denotes the realisation of the vector of standard normal variables $\mathbf{U} = \{U_1,\ldots,U_{N_r}\}$.

Figure 3.4 illustrates how the MPP-based methods work. Given a function $Y(\bullet)$ with random variables $\mathbf{X} = \{X_1, X_2\}$, these two random variables are transformed to

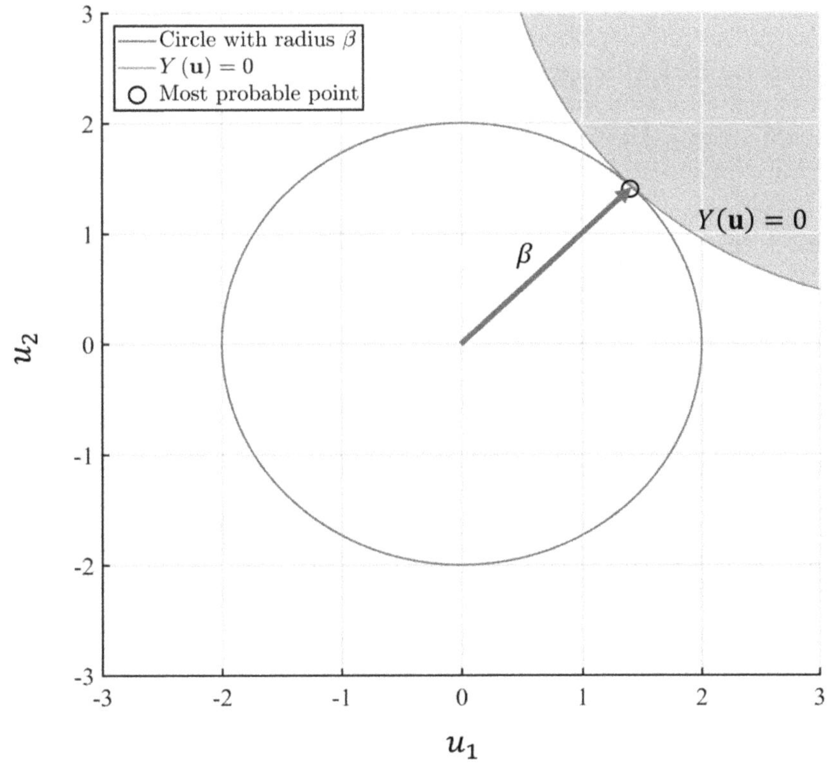

Figure 3.4. The most probable point in the normalised u-space for an arbitrary function $Y(\bullet)$ with two random variables X_1 and X_2. The symbol $\mathbf{u}$ denotes the realisations of $\mathbf{U}$; $\mathbf{U}$ denotes the normalised vector of random variables $\mathbf{X}$.

$\mathbf{U} = \{U_1, U_2\}$ using the transformation shown in equation (3.11). This brings them to a new u-space with a common centre point. The point on the line $Y(\mathbf{u}) = 0$ that has the shortest distance from the centre is known as the MPP, where $\mathbf{u} = \{u_1, u_2\}$ are realisations of $\mathbf{U} = \{U_1, U_2\}$, and the Euclidean distance between these two points corresponds to the reliability index β. This allows the original probabilistic formulation of uncertainty to be replaced by an approximated formulation that searches for the most probable point by computing the shortest distance between the origin and $Y(\mathbf{u}) = 0$ to find the β of the current design point.

Accordingly, for a given point $\mathbf{X}$, β can be found from the optimisation problem:

$$\begin{aligned} &\text{minimise: } \beta = \sqrt{\mathbf{u}^{\mathrm{T}}\mathbf{u}}, \\ &\text{subject to: } G(\mathbf{u}) = 0, \end{aligned} \tag{3.12}$$

where $\mathbf{u} = \{u_1,\ldots,u_{N_r}\}$ and $G(\cdot)$ is the constraint function in (3.1). The search for the MPP is also iterative and can be realised in three broadly different ways: the double-loop, decoupled-loop, and single-loop techniques, as illustrated in figure 3.5.

The FORM has been considered the mainstream classical uncertainty propagation method for RBDO since the 1980s [34], and further extensions have been made since the early 1990s [35, 36]. The SORM can be interpreted as an extension of the FORM using second-order approximation. This means that it uses the second-order partial derivatives of the function at the MPP [37, 38]. Since it has additional information, the SORM is generally more accurate. However, the use of second-order derivatives results in lower robustness compared to the FORM. Depending on the application, the FORM can be more desirable than the SORM, especially when the derivative information is numerically computed from *finite element method* (FEM) models.

Although the FORM and the SORM significantly reduce the computational complexity by transforming the uncertainty evaluation problem from the nested

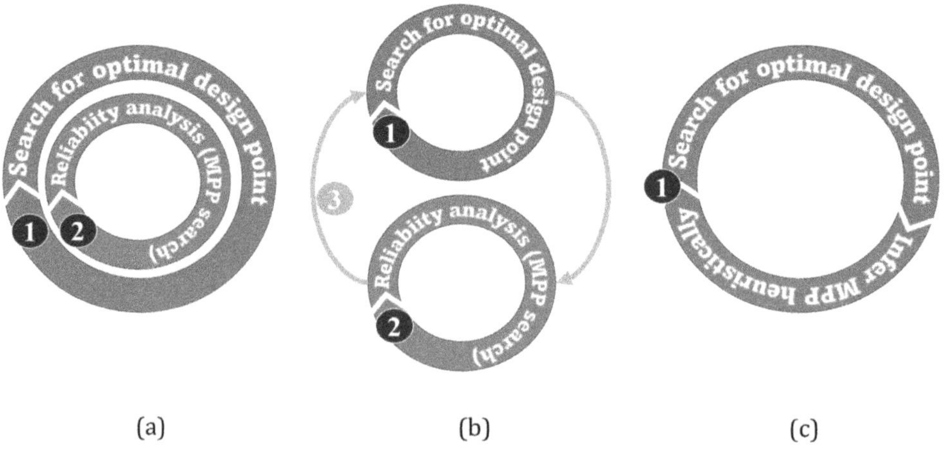

Figure 3.5. Different ways to deploy MPP-based RBDO methods: (a) represents a double-loop approach, (b) represents a decoupled-loop approach, and (c) represents a single-loop approach.

integral problem, they have several shortcomings. First, they may reduce the accuracy of the reliability estimation [39–42]. For example, [43] reports under- and overestimation of the reliability for concave and convex functions in the neighbourhood of the MPPs, respectively, which could potentially lead to unreliable designs. Figure 3.4, for instance, shows the MPP in the vicinity of a concave function [43]. Although the study went on to propose an MPP-based dimension reduction method that addressed this shortcoming, the results showed that it could only minimise the severity of under- or overestimations but not eliminate them.

Second, the convergence becomes poor in the search for MPPs representing high system reliability [23] and the possible existence of non-unique MPPs may lead to inaccurate reliability analysis [44]. Various attempts have been made to improve MPP-based uncertainty analysis; however, the whole class of techniques depends on finding the MPP reliably and efficiently. Therefore, it is still challenging to perform uncertainty analysis for some highly nonlinear problems containing highly skewed random variables.

3.2.1.1 The performance measure approach

Referring again to figure 3.5, the search for the MPP can be achieved via some innovative numerical methods, each of which achieves a different level of accuracy and computational efficiency. The double-loop strategy shown in figure 3.5(a) was first developed using the *reliability index approach* (RIA) [36] with Karush–Kuhn–Tucker optimisation conditions [45]. A sequential approximate programming strategy, known as the *performance measure approach* (PMA) [20, 46], was then proposed to replace the original problem of finding the reliability index in the RIA with a new formulation. The RBDO problem is represented using the PMA by equation (3.13). The target reliability is evaluated using inverse reliability analysis. In this analysis, the MPP is the failure point that corresponds to the lowest performance level that satisfies β_i, which is represented by (3.14).

$$\begin{aligned} \text{minimise:} \quad & C(\boldsymbol{d}), \\ \text{subject to:} \quad & G_i(\boldsymbol{d}, \boldsymbol{X}) \geqslant 0, \quad \text{for } i = 1,\ldots,m \\ & \boldsymbol{d}^L \leqslant \boldsymbol{d} \leqslant \boldsymbol{d}^U \end{aligned} \tag{3.13}$$

$$\begin{aligned} \text{minimise:} \quad & G_i(\boldsymbol{u}), \\ \text{subject to:} \quad & \|\, \boldsymbol{u} \,\| = \beta_i\text{'} \end{aligned} \tag{3.14}$$

where $G_i(\bullet)$ is the ith performance measure function that corresponds to the target reliability index β_i, $\boldsymbol{d}$ is the vector of design variables with lower bounds $\boldsymbol{d}^L$ and upper bounds $\boldsymbol{d}^U$, and $\boldsymbol{X}$ is the vector of random variables.

The reliability problem in (3.14) can be solved using algorithms such as the *advanced mean value* (AMV), the *conjugate mean value* (CMV), and the *hybrid mean value* (HMV) [46]. In general, the HMV method is considered the superior method, as it adaptively combines the AMV and CMV methods for a more robust MPP search [13, 47].

3.2.1.2 Sequential optimisation and reliability assessment

To further improve the computational efficiency, the PMA formulation was then inverted [48], and the MPP search was decoupled from the design optimisation loop as shown in figure 3.5(b). This method is widely known as *sequential optimisation and reliability assessment* (SORA) [49]. In SORA, deterministic optimisation is carried out in each cycle. The optimisation is shifted to a feasible region by changing the design vector **d** based on the MPP obtained in a previous cycle. The optimisation problem for SORA is represented by (3.15), where k is the current cycle of iteration and **s** is the shift vector that can be obtained from $\mathbf{s}^{k+1} = \mathbf{d}^k - \mathbf{X}^k_{MPP}$. The MPPs denoted by $\mathbf{X}_{MPP}$ in (3.15) are obtained using the inverse reliability strategy [48].

$$\begin{aligned} &\text{minimise:} \quad C(\mathbf{d}^k), \\ &\text{subject to:} \quad G_i(\mathbf{d}^k - \mathbf{s}^{k-1}, \mathbf{X}^{k-1}_{MPP}) \geqslant 0, \quad \text{for } i = 1,\ldots,m \\ &\qquad\qquad\quad \mathbf{d}^L \leqslant \mathbf{d} \leqslant \mathbf{d}^U. \end{aligned} \tag{3.15}$$

3.2.1.3 The single-loop approach

The *single-loop approach* (SLA) was introduced in [50, 51] to replace the reliability constraints with approximated deterministic constraints using the normalised gradient vector of the individual performance function $\boldsymbol{\alpha}_i = \{\alpha_1,\ldots,\alpha_{N_c}\}$ and the target reliability index β_i^t for $i = 1,\ldots,N_c$. With reference to figure 3.5(c), the SLA introduces no additional MPP search iterations or loops within the design optimisation iteration; consequently, it substantially improves the computational efficiency of the MPP search to a level comparable to that of the deterministic optimisation. Unfortunately, this is a 'pseudo-MPP' search which is not always guaranteed to satisfy nonlinear reliability constraints; thus, it leads to poorer convergence than those of the PMA and the SORA approaches.

The modern SLA [52] is represented by equation (3.16).

$$\begin{aligned} &\text{minimise:} && C(\mathbf{d}^k), , \\ &\text{subject to:} && G_i(\mathbf{d}^k, \mathbf{X}^k) \geqslant 0, \quad \text{for } i = 1, \ldots, m \\ &&& \mathbf{d}^L \leqslant \mathbf{d} \leqslant \mathbf{d}^U \\ &&& \mathbf{X}_i^k = \boldsymbol{\mu}_X^k - \boldsymbol{\alpha}_i^k \boldsymbol{\sigma}_X \beta_i^t \\ &&& \boldsymbol{\alpha}_i^k = \frac{\boldsymbol{\sigma}_X \nabla_X G_i(\mathbf{d}^k, \mathbf{X}_i^{k-1})}{\| \boldsymbol{\sigma}_X \nabla_X G_i(\mathbf{d}^k, \mathbf{X}_i^{k-1}) \|} \end{aligned} \tag{3.16}$$

where k is the current iteration cycle, $\mathbf{X}_i^k$ is the vector of the approximated MPP of the ith performance function, $\boldsymbol{\mu}_X^k$ is the vector of mean values, and $\boldsymbol{\sigma}_X$ is the vector of standard deviations.

Additional incremental improvements have been introduced for FORM-based uncertainty analysis in design optimisation over the last decade in order to find the MPP more efficiently and accurately [4, 53, 54].

3.2.2 The Monte Carlo method

The Monte Carlo method was presented in section 2.3 as one of the methods advocated for use by the GUM for uncertainty propagation of systems that cannot be adequately addressed by first-order linearisation [55]. It is used to simulate many independent realisations of the input quantities using the *a priori* probability distributions of the input quantities. The distributions are then propagated through $Y(\cdot)$ to find the output probability distribution from a histogram. Thus, the distribution of the output quantity can be reliably estimated if a sufficient number of simulations or trials is used. The resultant distribution can then be used to perform reliability and robustness assessments of the output Y (see figure 3.6).

However, the use of the MC method within an iterative procedure leads to two significant problems. The first problem is the lengthy computational time. This issue is very well documented in the engineering design optimisation literature [8, 12, 23, 56]. The second problem, however, is a subtler one. The MC method produces some simulation noise that can affect the stability or convergence of the iterative search/optimisation procedure. Both these problems are elaborated below by the use of examples.

First, on the matter of computational time, consider a lighting retrofit *measurement and verification* (M&V) project [57–59] in which a large number of inefficient lamps were replaced by energy-efficient ones. The M&V methodology is used to quantify the savings realised over several years using measurement instruments such as surveys and on-site energy meters. The uncertainty of the reported savings must fall within a given set of bounds to be eligible for the incentive programs of the funding body. Increasing the sample size would reduce the estimation uncertainty at the disadvantage of higher cost. Therefore, the goal is to design an optimal annual sampling/measurement plan in a manner which satisfies the uncertainty requirements at a minimal cost. In the design stage, the energy saving Y is computed at each time step as:

$$Y = aX_1X_2(X_3 - 1), \tag{3.17}$$

where a is the number of lamps retrofitted.

One of the uncertainty sources, namely the lamp survival X_1, is beta distributed, while the others, such as the mean annual energy usage X_2 and the ratio of power

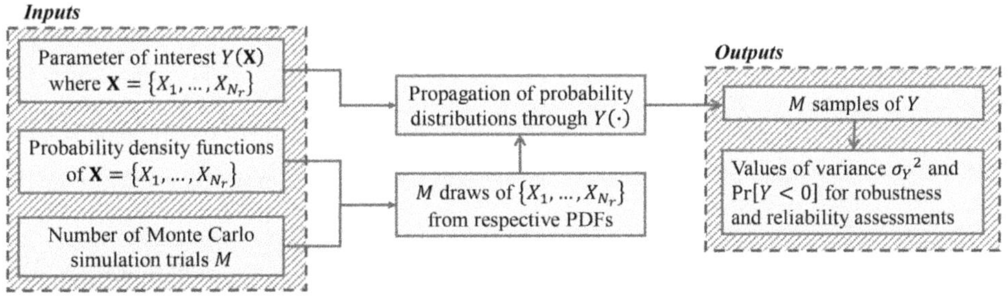

Figure 3.6. Framework of the Monte Carlo method for robustness and reliability assessments.

consumption X_3, are normally distributed. In one study, the *genetic algorithm* (GA) was used with integer linear programming [57] to design an appropriate sampling plan so that the M&V project results could satisfy the uncertainty requirement at the minimum sample size (cost). On a 64-bit Intel Core i7 quad-core CPU with 8 GB of memory running Linux Ubuntu 16.04 and Python 3.3 (*Numpy*), the use of the genetic algorithm with an MC trial size of 10^7, fifty individuals, and evolution over 50 generations took 320 min.

The computational speed becomes even more critical in the design optimisation of modern and complex engineering systems, e.g. structural, mechanical, and even electronic systems, in which the nonlinear model output (3.17) is replaced by the output from a model built using *finite-element* (FE) modeling [60]. FE models help to predict the dynamic interactions of a static or dynamic system and to compute the value of the parameter of interest conforming to the variable input parameters. In [61], for instance, one execution of the structural finite element model could take hours to complete, in which case the generation of 10^6 samples would have been practically impossible. Moreover, the computational burden increases exponentially when uncertainty evaluation has to be carried out iteratively, thus making design optimisation all the more impractical.

Second, consider the problem of optimiser stability. Optimisers can find solutions that are close to (while not violating) a specified coverage interval. The intersimulation variation between MC realisations means that the MC method can create a design that adheres to the constraints most of the time, but it can occasionally generate outliers that violate the constraints. Figure 3.7 illustrates this phenomenon. Let us assume that point O should fall within the 96th percentile. The MC method sometimes failed to satisfy this constraint (in MC realisation 1) due to simulation noise. When a new simulation was performed (in MC realisation 2), it seemed to fully satisfy the constraint. These 'false positives' or 'false negatives' happen rarely. However, if they occur, the logical consistency of the optimisation search is broken by the MC simulation noise. The optimiser often terminates while failing to

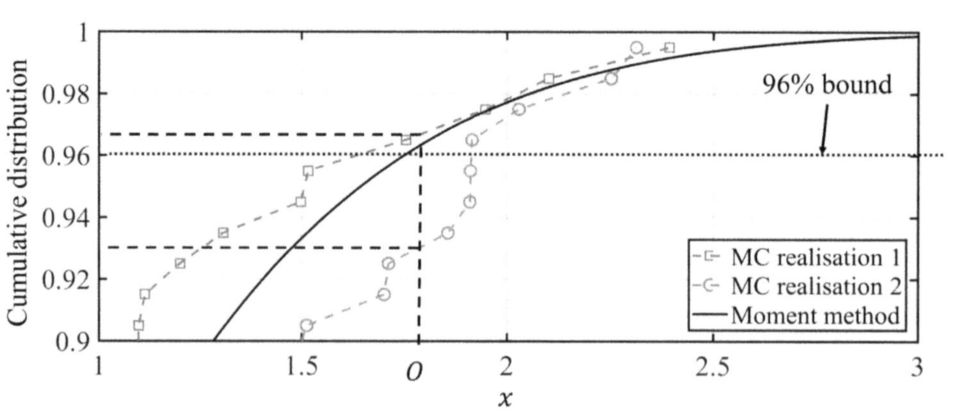

Figure 3.7. The cumulative distribution of a random variable X following the standard normal distribution obtained using the moment method (solid line) and the MC method (realisations 1 and 2), where x denotes the realisations of X and point O should cover less than the 96th percentile.

determine the reason for the constraint violation that randomly occurs. Unless the complete MC realisation is kept for a post-mortem inspection, the user is left with no indication as to why the optimisation algorithm randomly fails. The frequency of these occurrences increases with the complexity of the sampling plan and the target reliability (or confidence). In the worst-case scenario, some of the non-viable solutions can survive the optimisation process as viable ones due to the MC noise.

Alternatively, the frequency of occurrence may be reduced through the use of a larger MC trial size. However, doing so results in a significant increase in computation load without fundamentally addressing the root cause. Furthermore, only a marginal benefit may be derived from an increased trial size, since most of the additional trials are located in the high-mass regions of the PDF and not in the tails of the distribution which are of the main interest for reliability analysis. Alternatively, the MC method can be employed with importance sampling [62] for a marginal reduction in the required sample size and the simultaneous generation of more samples in the tail sections of the output PDF for better optimiser stability. In importance sampling, it is often necessary to find a sampling (or proposal) PDF with a 'reasonably small' variance. This is often a tricky task, and a wrong choice of sampling distribution has a detrimental effect on stability and accuracy [63, 64].

As a comparison, the method of uncertainty evaluation using moments (section 2.5) was also implemented (solid line) and this showed that it can satisfy this requirement consistently.

Due to its poor computational efficiency, poor constraint-adherence consistency throughout the optimisation process, and the availability of alternative uncertainty evaluation methods such as moment-based approaches, the use of the MC method (even with importance sampling) is less desirable for engineering design optimisation in which repeated uncertainty evaluation is required.

3.2.3 Metamodels with direct sampling methods

In a different line of development in the recent design optimisation literature, *metamodels* (or surrogate models) are employed to approximate the dynamic behaviour of output–inputs to improve the computational efficiency when implicit constraint and performance functions are evaluated using FE models [65]. This adds an additional step, which builds a metamodel of functions, between Steps 1 and 2 of the probabilistic optimisation framework shown in figure 3.1(b).

Metamodel-based probabilistic optimisations can be further categorised into global and local approximations. The former approach builds the metamodel for the entire design region before the optimisation process begins. The latter approach builds a new metamodel at every design point. Consequently, the global modelling approach is typically more computationally efficient, while the local modelling approach gives a more accurate representation of the functions [25]. Various innovative sampling schemes for metamodelling have been proposed to find the best trade-off between the accuracy and the efficiency of building metamodels. These are techniques such as Latin hypercube sampling [66], constraint boundary sampling [67], local adaptive sampling [24], sequential sampling [68], and selective interaction sampling [69].

The models that are normally used for metamodelling include: multivariate polynomials [69], polynomial chaos expansion [70], the radial basis function [27], artificial neural networks [26], and the Kriging method [28, 71]. The radial basis function provides an accurate global metamodel of functions, but it has the tendency to overfit, which leads to larger errors for optimisation under uncertainty [72]. Artificial neural networks are suitable for highly nonlinear and complex problems, but the computational expense of training a neural network is usually high [73]. Polynomial chaos expansion, on the other hand, is considered to be accurate, and the high-order moments of polynomial models can be analytically computed from the orthogonal polynomials [71]. As a full analytic method, polynomial chaos has been used in different niche areas for uncertainty propagation [74–77]. However, its computational efficiency and complexity heavily depend on the design of experiments, the optimal basis polynomial, the choice of that polynomial's degree, and the convergence [71].

Some of the well-known polynomial metamodelling methods are the polynomial dimensional decomposition [78], polynomial support vector machine [79, 80], high-order polynomial approximation [81], second-order polynomial using the least-squares approximation [82], and polynomial genetic programming [83, 84].

It is crucial to note that the development and assessment of effective sampling and polynomial modelling techniques are outside the scope of this book. This book focuses on the analytical computation of high-order moments of the multivariate polynomial models and finding the PDF of the corresponding moments. The effectiveness of the developed techniques will be demonstrated primarily by application to probabilistic engineering design optimisation, in which the choice of sampling and polynomial modelling technique is at the users' discretion.

This book embarks on finding the high-order moments analytically for all multivariate polynomials of arbitrary order in section 4.1. The reasons are discussed below.

Regarding the use of polynomials, the GUM has a dedicated section on the representation of polynomials [85], stating that 'polynomials are sometimes thought to be of very limited use because of problems of numerical reliability. In fact, it is their representation (that is parametrization) in terms of the monomial basis functions that leads to such problems, rather than polynomials *per se*' [85]. Thus, while it justifies the use of polynomials, it cautions that the choice of the polynomials could lead to subsequent difficulty when they are expressed as monomials.

Regarding the use of multivariate polynomials, the comparative studies [86, 87] recommend multivariate polynomials as the preferred metamodel among all the metamodelling techniques, as they have a good balance of accuracy, robustness, and computational efficiency. The use of multivariate polynomials is not a restrictive assumption, because the Stone–Weierstrass theorem [88] states that every continuous function defined in a compact space can be uniformly approximated as closely as desired by a polynomial function.

The subsequent subsections describe the commonly used multivariate polynomials.

3.2.3.1 Low-order polynomials

The use of low-order polynomials up to their second order is one of the most efficient and commonly used techniques for metamodelling in engineering design, and it is commonly known as the *response surface methodology* (RSM) [73, 89–91]. The least-squares approach [82] is commonly used to find the polynomial coefficients. This approach includes many variations such as the recursive least squares, partial least squares, weighted least squares, segmented least squares, and moving least squares [92].

Low-order polynomials struggle to accurately approximate moderate to highly nonlinear functions over a large region [73, 93]. This limits their use in global approximations such as the estimation of reliability or robustness for complex, nonlinear systems. Nonetheless, their efficiency makes them suitable for some types of local approximation, as shown in [91].

3.2.3.2 The Taylor series expansion

The Taylor series is a well-known power series expansion [94]. It is commonly used to approximate a continuous function $Y(X)$ by expanding the series to the desired order, where the choice of higher-order approximations generally results in a more accurate representation of $Y(X)$. The univariate Taylor series expansion of $Y(X)$ around a point a is obtained by taking the sum of its derivatives, as shown in expression (3.18).

$$\sum_{n=0}^{\infty} \frac{Y^{(n)}(a)}{n!}(X - a)^n \tag{3.18}$$

In (3.18), $Y^{(n)}(a)$ is the nth derivative of $Y(X)$ at the point a.

In probabilistic analyses, the series is normally expanded around the expected value (or mean) of the random variable $\boldsymbol{X}$, i.e. $a = E[\boldsymbol{X}]$. For multivariate performance functions, a multivariate Taylor series [95] may be more applicable.

3.2.3.3 High-order approximation using Chebyshev polynomials

Historically, high-order polynomials have not been employed, as they have been shown to result in ill-conditioned systems of equations that have significant differences between the approximated and true limit state functions outside the domain of the sample points [96, 97]. However, in 2008, a three-step method [81] was proposed to overcome these limitations through the use of Chebyshev polynomials:

1. Identify the highest order of the polynomial using Chebyshev orthogonal polynomials [98].
2. Determine the number and type of mixed terms in the polynomial.
3. Obtain the coefficient values of the polynomial using singular value decomposition [99].

Software that implements this method together with a full example of its use are provided in [100]; the software is written in the MATLAB programming language [101].

3.2.3.4 Polynomial genetic programming

Genetic programming is an approach inspired by biological evolution. An evolutionary algorithm is a set of model parameters that evolve to optimise the objective function. The use of genetic programming for metamodelling was first reported in [102] in 1998, followed by a 2001 paper [103] that extended it by using polynomials as the node functions in genetic programming.

In 2005, Yeun *et al* [83] introduced a more accurate and efficient polynomial genetic programming algorithm for response surface modelling and went on to demonstrate its reliable modelling capabilities for highly nonlinear functions. Their method was subsequently validated for use in the RBDO of structural systems in combination with MPP-based reliability analysis [84].

A decade later, in 2015, a symbolic genetic programming algorithm was presented [104] that took advantage of the availability of a free and open source program.

3.2.3.5 Polynomial support vector machines

Support vector machines (SVMs) constitute a classical machine learning approach with widespread applications in pattern recognition [105, 106]. A variation of the SVM is used as a metamodelling tool [80, 107] and has been shown to outperform other non-polynomial metamodels such as the Kriging [28] and radial basis function [26] approaches for RBDO applications.

3.2.3.6 Polynomial dimensional decomposition

Dimensional decomposition partitions a multivariate function into a finite sum of simpler component functions with respect to the input variables [108]. Taking advantage of this strategy, [78] proposed a metamodelling approach using the classical orthogonal polynomial expansion [109] with dimension reduction integration to calculate the coefficients of the expansions. Although the statistical moment can be analytically calculated from orthogonal polynomials, [110] reports that the convergence of the higher-order moments is not guaranteed using this approach. This is the same limitation as that encountered by the polynomial chaos expansion method [111].

To address this limitation, chapter 4 introduces an analytical computation method for higher-order moments for any polynomial expansion.

3.2.4 Moment-based methods for RBDO

Methods based on moments proceed by first calculating the high-order statistical moments of the output. While the variance for robustness assessment is computed from the first two moments, the PDF of the output for reliability assessment is approximated using a parametric distribution-fitting algorithm. The use of moments for uncertainty evaluation was previously presented in chapter 2. This section reviews the theoretical basis and the viability of the moment method within the context of reliability analysis.

The ith-order raw moment $\mathrm{E}[X^i]$ of a random variable X is defined as:

$$\mathrm{E}[X^i] = \int x^i f_X(x)\mathrm{d}x, \tag{3.19}$$

where $f_X(\cdot)$ denotes the PDF of X and x denotes the realisation of X.

The use of known moments to determine the corresponding underlying probability distribution is known as the *moment problem* [112]. The early studies in this field were done in the first half of the 20th century. They were mainly devoted to the existence conditions and uniqueness of the solutions. These classical investigations, summarised in [112, 113], considered cases in which a complete and possibly infinite sequence of moments was available. In practice, however, finding the complete moment sequence is not achievable except in certain special cases. Therefore, it is crucial to extend the findings of the traditional moment problem to the *truncated moment problem*, which considers a finite sequence of moments.

Recent findings in statistics have established the practical usefulness of solving the truncated moment problem. Instead of determining conditions that guarantee the uniqueness of a solution (or the probability distribution), attention has shifted to finding the tightest bounds for an appropriate probability distribution space in which all valid solutions of a given moment sequence can be found. This, in turn, quantifies the uncertainty of the potential distribution that can be obtained from the given set of moments.

Intuitively, this implies that the distributions must be like each other to some extent in order to generate identical first few moments. Furthermore, a longer moment sequence is indicative of tighter bounds; hence, there is potentially a smaller variation between different distributions obtained from this same set of moments. The types of bounds in the truncated moment problem and their properties are discussed in great detail in [114, 115]. They can be categorised as being of a global type [116], in which the largest discrepancies between all possible solutions over the full domain are sought, or of a local type [114, 115, 117], in which only the largest discrepancy at a particular point of interest is sought.

The global-type bounds tend to be more difficult to find because the information encoded in the moments concentrates on the tails of the distributions [114]. Furthermore, only the local-type bounds are required for reliability estimation [114, 115]. Note that this investigation excludes local-type bounds, such as the Markov/Chebyshev inequality, Chernoff's bound [118], the moment bound [119], and the fractional moment bound [120], as these bounds rely on moment-generating functions that assume the availability of a complete sequence of moments.

The findings reported in [114, 115] demonstrated that inferring the tails of a distribution from a finite sequence of moments is both robust and reliable with quantifiable uncertainty. Since the evaluation of reliability analysis concerns only the upper and sometimes the lower tails, the truncated moment method is well suited for the task. For example, Lindsay and Basak [114] showed that distribution fitting using n moments exhibited a good asymptotic approximation to the CDF tails with a worst-case convergence rate of x^{-N_m}, where N_m denotes the number of moments, in the allowable distribution space. In addition, Racz *et al* [115] reported procedures for estimating the *local distribution bounds* that quantify the largest possible discrepancy of all possible distributions based on the finite moment sequence at a particular point of interest. Both studies concluded that the bounds would improve asymptotically as a result of increasing: (1) the number N_m of known moments; or (2) the distance between x and the mean, regardless of the distribution (provided that it exists).

To illustrate this point, figure 3.8(a) shows the upper and lower bounds of the CDFs of X, denoted by $F_X(x)$, where x is the realisation of X, for which the first four moments (dashed blue line) and first eight moments (dotted red line) are identical to those of the standard normal distribution. In other words, figure 3.8(a) illustrates the theoretical worst-case deviations of the possible CDFs that can be obtained from these moments in comparison to the actual CDF of the standard normal distribution. It is important to emphasise that the obtained bounds are independent of the distribution-fitting technique used. Figure 3.8(b), on the other hand, shows the bound gap (the difference between the upper and lower bound) of X, denoted by $\rho_X(x)$. It is evident from the figures that the permissible distribution space shrinks when: (1) the number of the moments N_m increases; or (2) x moves further from the mean.

These findings demonstrate the excellent reliability of the truncated moment method, provided that the moments of a sufficiently large order N_m are available. Unfortunately, it is uncommon to find the use of $N_m > 4$ moments in the scientific literature. This is regrettable, since there is a smaller admissible distribution space for $N_m > 4$. For example, the dashed line in figure 3.8(b) shows that the admissible distribution space at $x \approx 3.75$ for $N_m = 4$ moments is larger than that of $N_m = 8$ moments by the factor of ten. Therefore, the uncommon use of more than four moments in the published literature may lead to the misguided perception that the moment approach is somewhat unreliable.

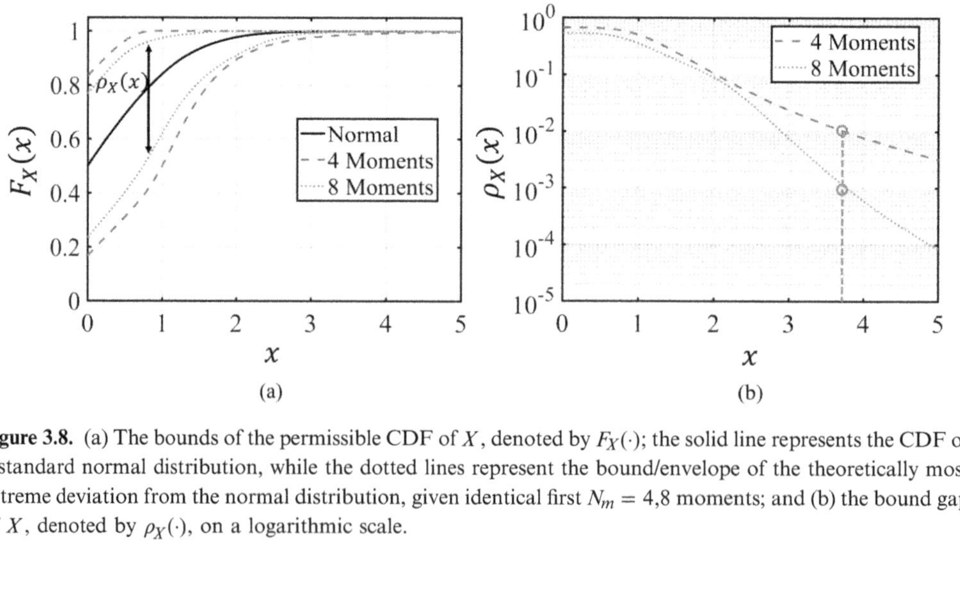

Figure 3.8. (a) The bounds of the permissible CDF of X, denoted by $F_X(\cdot)$; the solid line represents the CDF of a standard normal distribution, while the dotted lines represent the bound/envelope of the theoretically most extreme deviation from the normal distribution, given identical first $N_m = 4,8$ moments; and (b) the bound gap of X, denoted by $\rho_X(\cdot)$, on a logarithmic scale.

3.3 Reliability-based robust design optimisation

Formally, reliability-based robust design optimisation or RBRDO [12, 13] is a unified framework that combines: (a) the reliability-based design optimisation (RBDO), which optimises the design objectives for a given set of probabilistic (or reliability) constraints; and (b) the robust design optimisation (RDO), which increases the robustness of a designed system by minimising the sensitivity of the design objectives to process variabilities. The integration of RDO and RBDO offers a complete solution that assesses the best compromise between cost, reliability, and robustness. By incorporating both formulations, i.e. (3.2) and (3.9), a typical RBRDO problem can be mathematically represented as follows:

$$\begin{aligned} &\text{minimise:} && C\left(\mu_K, {\sigma_K}^2\right), \\ &\text{subject to:} && \Pr[G_i(\mathbf{d}, \mathbf{X}) \leqslant 0] > \Phi\left(\beta_i\right) \text{ for } i = 1, \ldots, N_{c,} \\ &\text{where:} && \mathbf{d}^L \leqslant \mathbf{d} \leqslant \mathbf{d}^U. \end{aligned} \tag{3.20}$$

Although this method is predominantly used in the field of structural and mechanical design, in recent years, it has become increasingly popular in other fields of engineering, such as magnetics, manufacturing, microelectronics, micromachining, etc [121–123] due to the increasingly competitive market conditions, stringent safety requirements, and the ability of this method to deliver designs that are insensitive to uncontrollable variations [122].

It should be noted that while the incorporation of robustness analysis into the RBDO framework could lead to better-quality designs, it incurs a high computational load/time. Therefore, such frameworks are often characterised by an inherent trade-off between accuracy and computational complexity.

3.4 Summary

Reliability-based robust design optimisation is a framework that combines reliability-based design optimisation and robust design optimisation where RBDO optimises design objectives based on probabilistic constraints, while RDO increases the robustness of a design by minimising its sensitivity to process variations. RBRDO offers a compromise between cost, reliability, and robustness and is widely used in various fields of engineering, such as structural and mechanical engineering, magnetics, manufacturing, microelectronics, and micromachining. However, incorporating robustness analysis into RBDO incurs a high computational cost, resulting in a trade-off between accuracy and computational complexity.

References

[1] Fesanghary M, Mahdavi M, Minary-Jolandan M and Alizadeh Y 2008 Hybridizing harmony search algorithm with sequential quadratic programming for engineering optimization problems *Comput. Meth. Appl. Mech. Eng.* **197** 3080–91

[2] Paul S, Rajan A, Chang J, Kuang Y C and Ooi M P-L 2018 Parametric design analysis of magnetic sensor based on model order reduction and reliability-based design optimization *IEEE Trans. Magn.* **54** 1–4
[3] Rajan A, Luo F J, Kuang Y C, Bai Y and Ooi M P-L 2020 Reliability-based design optimisation of structural systems using high-order analytical moments *Struct. Saf.* **86** 101970
[4] Okasha N M 2016 An improved weighted average simulation approach for solving reliability-based analysis and design optimization problems *Struct. Saf.* **60** 47–55
[5] Allen M, Raulli M, Maute K and Frangopol D M 2004 Reliability-based analysis and design optimization of electrostatically actuated MEMS *Comput. Struct.* **82** 1007–20
[6] Nocedal J and Wright S J 2006 Quadratic programming *Numerical Optimization* (New York, NY: Springer) pp 448–92
[7] Deb K, Pratap A, Agarwal S and Meyarivan T 2002 A fast and elitist multiobjective genetic algorithm: NSGA-II *IEEE Trans. Evol. Comput.* **6** 182–97
[8] Valdebenito M A and Schuller G I 2010 A survey on approaches for reliability-based optimization *Struct. Multidiscip. Optim.* **42** 645–63
[9] Helton J C and Davis F J 2003 Latin hypercube sampling and the propagation of uncertainty in analyses of complex systems *Reliab. Eng. Syst. Saf.* **81** 23–69
[10] Park G-J, Lee T-H, Lee K H and Hwang K-H 2006 Robust design: an overview *AIAA J.* **44** 181–91
[11] Zang C, Friswell M and Mottershead J 2005 A review of robust optimal design and its application in dynamics *Comput. Struct.* **83** 315–26
[12] Lee I, Choi K K, Du L and Gorsich D 2008 Dimension reduction method for reliability-based robust design optimization *Comput. Struct.* **86** 1550–62
[13] Shahraki A F and Noorossana R 2014 Reliability-based robust design optimization: a general methodology using genetic algorithm *Comput. Ind. Eng.* **74** 199–207
[14] Youn B D, Choi K K and Yi K 2005 Performance moment integration (PMI) method for quality assessment in reliability-based robust design optimization *Mech. Based Des. Struct. Mach.* **33** 185–213
[15] Rahman S and Xu H 2004 A univariate dimension-reduction method for multi-dimensional integration in stochastic mechanics *Probab. Eng. Mech.* **19** 393–408
[16] Mourelatos Z P and Liang J 2006 A methodology for trading-off performance and robustness under uncertainty *J. Mech. Des.* **128** 856–63
[17] Yadav O P, Bhamare S S and Rathore A 2010 Reliability-based robust design optimization: a multi-objective framework using hybrid quality loss function *Qual. Reliab. Eng. Int.* **26** 27–41
[18] Youn B D, Choi K K, Yang R J and Gu L 2004 Reliability-based design optimization for crashworthiness of vehicle side impact *Struct. Multidiscip. Optim.* **26** 272–83
[19] Rosenblatt M 1952 Remarks on a multivariate transformation *Ann. Math. Stat.* **23** 470–2
[20] Youn B D, Choi K K and Park Y H 2003 Hybrid analysis method for reliability-based design optimization *J. Mech. Des.* **125** 221–32
[21] Ezzati G, Mammadov M and Kulkarni S 2015 A new reliability analysis method based on the conjugate gradient direction *Struct. Multidiscip. Optim.* **51** 89–98
[22] Goswami S, Ghosh S and Chakraborty S 2016 Reliability analysis of structures by iterative improved response surface method *Struct. Saf.* **60** 56–66

[23] Aoues Y and Chateauneuf A 2010 Benchmark study of numerical methods for reliability-based design optimization *Struct. Multidiscip. Optim.* **41** 277–94
[24] Chen Z, Qiu H, Gao L, Li X and Li P 2014 A local adaptive sampling method for reliability-based design optimization using Kriging model *Struct. Multidiscip. Optim.* **49** 401–16
[25] Li X, Qiu H, Chen Z, Gao L and Shao X 2016 A local Kriging approximation method using MPP for reliability-based design optimization *Comput. Struct.* **162** 102–15
[26] Cheng J and Li Q S 2008 Reliability analysis of structures using artificial neural network based genetic algorithms *Comput. Meth. Appl. Mech. Eng.* **197** 3742–50
[27] Dai H Z, Zhao W, Wang W and Cao Z G 2011 An improved radial basis function network for structural reliability analysis *J. Mech. Sci. Technol.* **25** 2151
[28] Ju B and Lee B 2008 Reliability-based design optimization using a moment method and a kriging metamodel *Eng. Optim.* **40** 421–38
[29] Low Y M 2013 A new distribution for fitting four moments and its applications to reliability analysis *Struct. Saf.* **42** 12–25
[30] Lu Z-H, Cai C-H and Zhao Y-G 2017 Structural reliability analysis including correlated random variables based on third-moment transformation *J. Struct. Eng.* **143** 04017067
[31] Lu Z-H, Hu D-Z and Zhao Y-G 2016 Second-order fourth-moment method for structural reliability *J. Eng. Mech.* **143** 06016010
[32] Zhang X and Pandey M D 2013 Structural reliability analysis based on the concepts of entropy, fractional moment and dimensional reduction method *Struct. Saf.* **43** 28–40
[33] Zhao Y-G, Zhang X-Y and Lu Z-H 2018 Complete monotonic expression of the fourth-moment normal transformation for structural reliability *Comput. Struct.* **196** 186–99
[34] Nikolaidis E and Burdisso R 1988 Reliability based optimization: a safety index approach *Comput. Struct.* **28** 781–8
[35] Enevoldsen I and Srensen J D 1994 Reliability-based optimization in structural engineering *Struct. Saf.* **15** 169–96
[36] Tu J, Choi K K and Park Y H 1999 A new study on reliability-based design optimization *J. Mech. Des.* **121** 557–64
[37] Haldar A and Mahadevan S 1999 *Probability, Reliability, and Statistical Methods in Engineering Design* (Hoboken, NJ: Wiley) https://www.wiley.com/en-be/Probability,+Reliability,+and+Statistical+Methods+in+Engineering+Design-p-9780471331193
[38] Melchers R E and Beck A T 2017 *Structural Reliability Analysis and Prediction* 3 (Hoboken, NJ: Wiley)
[39] Zhao Y-G and Ono T 2001 Moment methods for structural reliability *Struct. Saf.* **23** 47–75
[40] Lee S H and Kwak B M 2006 Response surface augmented moment method for efficient reliability analysis *Struct. Saf.* **28** 261–72
[41] Youn B D and Xi Z 2009 Reliability-based robust design optimization using the eigenvector dimension reduction (EDR) method *Struct. Multidiscip. Optim.* **37** 475–92
[42] Kang H Y and Kwak B M 2009 Application of maximum entropy principle for reliability-based design optimization *Struct. Multidiscip. Optim.* **38** 331–46
[43] Lee I, Choi K K and Gorsich D 2010 System reliability-based design optimization using the MPP -based dimension reduction method *Struct. Multidiscip. Optim.* **41** 823–39
[44] Der Kiureghian A and Dakessian T 1998 Multiple design points in first and second-order reliability *Struct. Saf.* **20** 37–49

[45] Kuschel N and Rackwitz R 1997 Two basic problems in reliability-based structural optimization *Math. Methods Oper. Res.* **46** 309–33
[46] Youn B D, Choi K K and Du L 2005 Enriched performance measure approach for reliability-based design optimization *AIAA J.* **43** 874–84
[47] Youn B D and Choi K K 2004 An investigation of nonlinearity of reliability-based design optimization approaches *J. Mech. Des.* **126** 403–11
[48] Du X, Sudjianto A and Chen W 2004 An integrated framework for optimization under uncertainty using inverse reliability strategy *J. Mech. Des.* **126** 562–70
[49] Du X and Chen W 2004 Sequential optimization and reliability assessment method for efficient probabilistic design *J. Mech. Des.* **126** 225–33
[50] Agarwal H, Mozumder C K, Renaud J E and Watson L T 2007 An inverse-measure-based unilevel architecture for reliability-based design optimization *Struct. Multidiscip. Optim.* **33** 217–27
[51] Shan S and Wang G G 2008 Reliable design space and complete single-loop reliability-based design optimization *Reliab. Eng. Syst. Saf.* **93** 1218–30
[52] Liang J, Mourelatos Z P and Tu J 2008 A single-loop method for reliability-based design optimisation *Int. J. Prod. Devel.* **5** 76–92
[53] Roudak M A, Shayanfar M A, Barkhordari M A and Karamloo M 2017 A robust approximation method for nonlinear cases of structural reliability analysis *Int. J. Mech. Sci.* **133** 11–20
[54] Keshtegar B and Hao P 2018 A hybrid descent mean value for accurate and efficient performance measure approach of reliability-based design optimization *Comput. Meth. Appl. Mech. Eng.* **336** 237–59
[55] Guides in Metrology Committee: JCGM 2008 *Supplement 1 – Propagation of distributions using a Monte Carlo method (2008)* (Sèvres: BIPM) https://www.bipm.org/en/committees/jc/jcgm/publications JCGM 101:2008
[56] Lagaros N D and Papadrakakis M 2007 Robust seismic design optimization of steel structures *Struct. Multidiscip. Optim.* **33** 457–69
[57] Carstens H, Xia X, Yadavalli S and Rajan A 2017 Efficient longitudinal population survival survey sampling for the measurement and verification of lighting retrofit projects *Energy Build.* **150** 163–76
[58] Carstens H, Xia X and Yadavalli S 2017 Low-cost energy meter calibration method for measurement and verification *Appl. Energy* **188** 563–75
[59] Carstens H, Xia X and Yadavalli S 2017 Efficient metering and surveying sampling designs in longitudinal measurement and verification for lighting retrofit *Energy Build.* **154** 430–47
[60] Hughes T J R 2012 *The Finite Element Method: Linear Static and Dynamic Finite Element Analysis* (Mineola, NY: Dover Publications Inc.) https://store.doverpublications.com/0486411818.html
[61] Paul S and Chang J 2017 Design and parametric study of the magnetic sensor for position detection in linear motor based on nonlinear parametric model order reduction *Sensors* **17** 1543
[62] Robert C P and Casella G 2004 *Monte Carlo Statistical Methods* (New York, NY: Springer)
[63] Engelund S and Rackwitz R 1993 A benchmark study on importance sampling techniques in structural reliability *Struct. Saf.* **12** 255–76
[64] Melchers R E 1989 Importance sampling in structural systems *Struct. Saf.* **6** 3–10

[65] Missoum S, Ramu P and Haftka R T 2007 A convex hull approach for the reliability-based design optimization of nonlinear transient dynamic problems *Comput. Meth. Appl. Mech. Eng.* **196** 2895–906

[66] Iman R L 2006 Latin hypercube sampling *Encyclopedia of Statistical Sciences* (Hoboken, NJ: Wiley)

[67] Lee T H and Jung J J 2008 A sampling technique enhancing accuracy and efficiency of metamodel-based RBDO: constraint boundary sampling *Comput. Struct.* **86** 1463–76

[68] Zhao L, Choi K K, Lee I and Du L 2010 *Response Surface Method Using Sequential Sampling for Reliability-Based Design Optimization ASME 2009 Int. Design Engineering Technical Confs. and Computers and Information in Engineering Conf.* (New York, NY: The American Society of Mechanical Engineers) 1171–81

[69] Youn B D and Choi K K 2004 A new response surface methodology for reliability-based design optimization *Comput. Struct.* **82** 241–56

[70] Hu C and Youn B D 2011 Adaptive-sparse polynomial chaos expansion for reliability analysis and design of complex engineering systems *Struct. Multidiscip. Optim.* **43** 419–42

[71] Echard B, Gayton N and Lemaire M 2011 AK-MCS: an active learning reliability method combining Kriging and Monte Carlo simulation *Struct. Saf.* **33** 145–54

[72] Jin R, Du X and Chen W 2003 The use of metamodeling techniques for optimization under uncertainty *Struct. Multidiscip. Optim.* **25** 99–116

[73] Simpson T W, Poplinski J D, Koch P N and Allen J K 2001 Metamodels for computer-based engineering design: survey and recommendations *Eng. Comput.* **17** 129–50

[74] Xiu D and Karniadakis G E 2002 The Wiener–Askey polynomial chaos for stochastic differential equations *SIAM J. Sci. Comput.* **24** 619–44

[75] Xiu D and Karniadakis G E 2003 Modeling uncertainty in flow simulations via generalized polynomial chaos *J. Comput. Phys.* **187** 137–67

[76] Najm H N 2009 Uncertainty quantification and polynomial chaos techniques in computational fluid dynamics *Annu. Rev. Fluid Mech.* **41** 35–52

[77] Eldred M and Burkardt J 2012 *Comparison of Non-Intrusive Polynomial Chaos and Stochastic Collocation Methods for Uncertainty Quantification 47th AIAA Aerospace Sciences Meeting including The New Horizons Forum and Aerospace Exposition* (Reston, VA: American Institute of Aeronautics and Astronautics)

[78] Rahman S 2008 A polynomial dimensional decomposition for stochastic computing *Int. J. Numer. Methods Eng.* **76** 2091–116

[79] Cortes C and Vapnik V 1995 Support-vector networks *Mach. Learn.* **20** 273–97

[80] Basudhar A and Missoum S 2008 Adaptive explicit decision functions for probabilistic design and optimization using support vector machines *Comput. Struct.* **86** 1904–17

[81] Gavin H P and Yau S C 2008 High-order limit state functions in the response surface method for structural reliability analysis *Struct. Saf.* **30** 162–79

[82] Khuri A I and Mukhopadhyay S 2010 Response surface methodology *Wiley Interdiscip. Rev. Comput. Stat.* **2** 128–49

[83] Yeun Y S, Yang Y S, Ruy W S and Kim B J 2005 Polynomial genetic programming for response surface modeling. Part 1: a methodology *Struct. Multidiscip. Optim.* **29** 19–34

[84] Yeun Y S, Kim B J, Yang Y S and Ruy W S 2005 Polynomial genetic programming for response surface modeling. Part 2: adaptive approximate models with probabilistic optimization problems *Struct. Multidiscip. Optim.* **29** 35–49

[85] Guides in Metrology Committee: JCGM 2020 *Guide to the Expression of Uncertainty in Measurement—Part 6: Developing and Using Measurement Models* (Sèvres: BIPM) https://www.bipm.org/en/committees/jc/jcgm/publications JCGM GUM-6:2020
[86] Zhao D and Xue D 2010 A comparative study of metamodeling methods considering sample quality merits *Struct. Multidiscip. Optim.* **42** 923–38
[87] Bucher C and Most T 2008 A comparison of approximate response functions in structural reliability analysis *Probab. Eng. Mech.* **23** 154–63
[88] Hewitt E and Zuckerman H S 1959 Approximation by polynomials with integral coefficients, a reformulation of the Stone–Weierstrass theorem *Duke Math. J.* **26** 305–24
[89] Gunst R F 1996 Response Surface Methodology: Process and Product Optimization Using Designed Experiments *Technometrics* **38** 284–6
[90] Mullur A A and Messac A 2006 Metamodeling using extended radial basis functions: a comparative approach *Eng. Comput.* **21** 203
[91] Mansour R and Olsson M 2016 Response surface single loop reliability-based design optimization with higher-order reliability assessment *Struct. Multidiscip. Optim.* **54** 63–79
[92] Zhang H, Guo C, Su X and Zhu C 2015 Measurement data fitting based on moving least squares method *Math. Prob. Eng.* **2015** 195023
[93] Hussain M F, Barton R R and Joshi S B 2002 Metamodeling: radial basis functions, versus polynomials *Eur. J. Oper. Res.* **138** 142–54
[94] Lang S 2012 *A First Course in Calculus* (New York, NY: Springer)
[95] Neidinger R 2005 Directions for computing truncated multivariate Taylor series *Math. Comput.* **74** 321–40
[96] Olivi L 1980 Response surface methodology in risk analysis *Synthesis and Analysis Methods for Safety and Reliability Studies* (Boston, MA: Springer) pp 313–27
[97] Rajashekhar M R and Ellingwood B R 1993 A new look at the response surface approach for reliability analysis *Struct. Saf.* **12** 205–20
[98] Mason J C and Handscomb D C 2002 *Chebyshev Polynomials* (New York, NY: Chapman and Hall/CRC)
[99] Golub G H and Reinsch C 1970 Singular value decomposition and least squares solutions *Numer. Math.* **14** 403–20
[100] Gavin H P High Order Stochastic Response Surface Method—HOSRSM Duke University https://people.duke.edu/~hpgavin/hosrsm/ (accessed 7 September 2022)
[101] MATLAB version 9.1.0.441655 (R2016b) 2016 (Natick, MA: The MathWorks Inc.)
[102] Toropov V V and Alvarez L F 1998 Application of genetic programming and response surface methodology to optimization and inverse problems *Inverse Problems in Engineering Machanics Int. Symp. on Inverse Problems in Engineering Mechanics 1998 (ISIP '98) (Nagano, Japan)* M Tanaka and G S Dulikravich (Amsterdam: Elsevier) pp 551–60
[103] Nikolaev N Y and Iba H 2001 Regularization approach to inductive genetic programming *IEEE Trans. Evol. Comput.* **5** 359–75
[104] Searson D P 2015 GPTIPS 2: an open-source software platform for symbolic data mining *Handbook of Genetic Programming Applications* (Cham: Springer) pp 551–73
[105] Shawe-Taylor J and Cristianini N 2004 *Kernel Methods for Pattern Analysis* (Cambridge: Cambridge University Press)
[106] Tou J T and Gonzalez R C 1974 *Pattern Recognition Principles* (Reading, MA: Addison-Wesley)

[107] Clarke S M, Griebsch J H and Simpson T W 2005 Analysis of support vector regression for approximation of complex engineering analyses *J. Mech. Des.* **127** 1077–87
[108] Rabitz H and Ali 1999 General foundations of high-dimensional model representations *J. Math. Chem.* **25** 197–233
[109] Szegő G 1939 *Orthogonal Polynomials* 23 (Providence, RI: American Mathematical Society) https://bookstore.ams.org/coll-23
[110] Rahman S 2009 Statistical moments of polynomial dimensional decomposition *J. Eng. Mech.* **136** 923–7
[111] Xiu D and Karniadakis G E 2002 Modeling uncertainty in steady state diffusion problems via generalized polynomial chaos *Comput. Meth. Appl. Mech. Eng.* **191** 4927–48
[112] Aheizer N I 1965 *The Classical Moment Problem and Some Related Questions in Analysis* (Mineola, NY: Dover Publications) https://store.doverpublications.com/0486845559.html
[113] Shohat J A and Tamarkin J D 1943 *The Problem of Moments* vol 1 (Providence, RI: American Mathematical Society)
[114] Lindsay B G and Basak P 2000 Moments determine the tail of a distribution (but not much else) *Am. Stat.* **54** 248–51
[115] Racz S, Tari A and Telek M 2006 A moments based distribution bounding method *Math. Comput. Modell.* **43** 1367–82
[116] Tardella L 2001 A note on estimating the diameter of a truncated moment class *Stat. Probab. Lett.* **54** 115–24
[117] Royden H L 1953 Bounds on a distribution function when its first *n* moments are given *Ann. Math. Stat.* **24** 361–76
[118] Chernoff H 1952 A measure of asymptotic efficiency for tests of a hypothesis based on the sum of observations *Ann. Math. Stat.* **23** 493–507
[119] Philips T K and Nelson R 1995 The moment bound is tighter than Chernoff's bound for positive tail probabilities *Am. Stat.* **49** 175–8
[120] Goria M N and Tagliani A 2003 Bounds on the tail probability and absolute difference between two distributions *Commun. Stat. Theory Methods* **32** 519–32
[121] Wan Y, Huang H and Pecht M 2015 Thermal fatigue reliability analysis and structural optimization based on a robust method for microelectronics FBGA packages *IEEE Trans. Device Mater. Reliab.* **15** 206–13
[122] Ren Z, Zhang D and Koh C-S 2013 New reliability-based robust design optimization algorithms for electromagnetic devices utilizing worst case scenario approximation *IEEE Trans. Magn.* **49** 2137–40
[123] Jang J *et al* 2015 Reliability-based robust design optimization with kernel density estimation for electric power steering motor considering manufacturing uncertainties *IEEE Trans. Magn.* **51** 1–4

Melanie Po-Leen Ooi, Arvind Rajan, Ye Chow Kuang and Serge Demidenko

Chapter 4

Moment-based standard uncertainty in design optimisation

The use of high-order moments (those with large N_m) improves the estimation of probability distribution, as discussed in chapters 2 and 3. However, the extensive literature on this method has shown that the reliable computation of the high-order moments, especially $N_m \geqslant 4$, is numerically challenging. The truncated moment problem is an ongoing topic of active research, in which innovative methods are still being developed to compute high-order moments effectively.

While the MPP-based methods have been widely reported in the literature, the use of high-order ($N_m \geqslant 4$) moment-based methods for uncertainty or reliability analysis is unfortunately scarce. The first introduction of this approach was described in [1] in the early 1960s. Since then, more computationally efficient approximations of the high-order moments have been developed for engineering applications. For example, fewer samples for system response approximation based on second-order polynomial models were used in [2] for moment calculation; a numerical quadrature rule was proposed in [3], which was later coupled with the Kriging response surface method [4]; a dimension reduction strategy [5, 6] and orthogonal polynomial strategies [7] were introduced to find the high-order moments more efficiently; and some developments [8, 9] have advocated the use of moment-based methods by employing the third- and fourth-moment reliability index.

For the variance computation in moment-based robustness analysis discussed in section 3.1, *performance moment integration* (PMI) [10] and the univariate *dimension reduction method* (DRM) [11] are normally used. These improvements are strictly based on the first four moments and/or the use of numerical approximations, which are susceptible to sampling and approximation errors with increasing orders [12]. Furthermore, for reliability analysis, section 3.2 showed that accurately calculated moments result in a more accurate computation with increasing orders of moments.

doi:10.1088/978-0-7503-4931-4ch4

These studies favour the use of moment-based methods in reliability analysis on the condition that an accurate tool for moment calculation is available. In addition to these advantages, the confidence interval of the reliability estimation can also be quantified non-parametrically using the available exact moments. These characteristics are highly valuable to the design and regulation of safety-critical engineering systems, thus making the moment-based approach a viable candidate for reliability analysis.

As stated in the previous chapters, this book takes the approach of using just one line of approximation for uncertainty propagation through moments, which is to represent (or approximate) the output of interest using a polynomial function. This chapter presents an analytical moment evaluation framework for multivariate polynomials that employs a mathematical transform known as the Mellin transform [13] in section 4.1. It assumes that the function is already in the form of a polynomial. If this is not the case, section 4.1 presents a method that can be used to accurately estimate any given function using polynomials; the method is compatible with the analytical moment-based standard uncertainty evaluation presented here. Section 4.2 introduces an open-source toolbox that can be used to calculate the standard uncertainty using the moments based on the methodology in section 4.1, followed by some case studies of its application in section 4.3. Its incorporation into the probabilistic design optimisation framework is then presented in section 4.4.

4.1 The derivation of the analytical moments of multivariate polynomials

The biggest obstacle to analytically deriving high-order moments is the complex mathematical operations involving the application of Cauchy's integral theorem to evaluate Fox's H-function [14]. The use of the Mellin transform [13] to analytically calculate the standard uncertainty of a multivariate polynomial was first presented in 2013 [15]. This paper showed that the general form of the output PDF $f_Y(\cdot)$ after Mellin transformation can be expressed as an instance of Fox's H-function, provided that it is initially expressed as a multivariate polynomial. Once $f_Y(\cdot)$ is obtained, it can be used to perform reliability analysis in optimisation, and the high-order moments can be derived to perform robustness analysis.

Figure 4.1 presents a procedural analytical moment derivation framework of arbitrary order to guide the practical calculations. It takes advantage of the constraints imposed by the analytical moment derivation problem to simplify the H-function derivation. A mathematical proof of this simplification procedure is presented in section 4.1.1. It is then used in conjunction with a specially derived Mellin transform lookup table as a replacement for the complex mathematical derivations typically required by the analytical approach.

First, the output function $Y(\cdot)$ must be known and expressed as a multivariate polynomial function of the input variables $\mathbf{X} = \{X_1, \dots, X_{N_r}\}$. Next, the terms of $Y^i(\cdot)$ are expressed, with $i = 1, \dots, N_m$ denoting the desired order of the moments. The ith-order moment of Y is denoted by $\mathrm{E}[Y^i]$. Section 4.1.2 demonstrates that

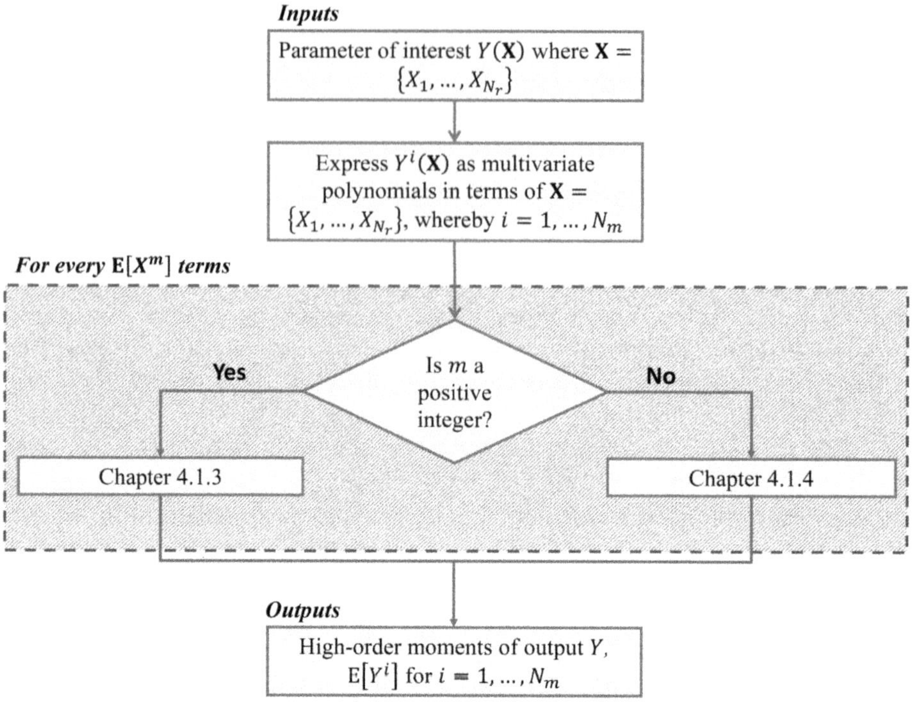

Figure 4.1. The analytical moment calculation method using the Mellin transform for multivariate polynomials.

moment evaluation does not require the computation of the PDF of Y and that it is reducible to the moment estimation of its monomials. Therefore, the problem of evaluating $\mathrm{E}[Y^i]$ is reduced to the problem of finding $\mathrm{E}[X_i^m]$, where $i = 1, \ldots, N_r$ and m is an integer. Section 4.1.3 goes on to demonstrate that the evaluation of $\mathrm{E}[X_i^m]$ can easily be accomplished using a Mellin transform lookup table accompanied by the formulations listed in sections 4.1.3 and 4.1.4, depending on the the sign of the integer m, i.e., whether m is positive or negative. The high-order moments $\mathrm{E}[Y^i]$ can finally be expressed in terms of the statistical parameters of the inputs.

4.1.1 The Mellin transform and the product of independent random variables

Let Y be represented by a set of independent and positive random variables X_i as shown in equation (4.1) with known probability density functions $f_{X_i}(\cdot)$. Consider that V is a type of random variable X_i that has a closed-form Mellin transform expression and therefore has a standard tabulated distribution; the PDF $f_Y(y)$ of output Y can then be deduced from equation (4.2), where $\mathcal{M}[\cdot]$ and $\mathcal{M}^{-1}[\cdot]$ are the Mellin transform and its inverse as defined by equations (4.3) and (4.4), respectively [16].

$$Y = \prod_i X_i. \tag{4.1}$$

$$f_Y(y) = \mathcal{M}^{-1}\left[\prod_i \mathcal{M}[f_V(v_i)]\right] \tag{4.2}$$

$$M_{v_i}(s) = \mathcal{M}\left[f_V(v_i)\right](s) = \int_0^\infty v_i^{s-1} f_{V_i}(v_i)\mathrm{d}v_i \tag{4.3}$$

$$\mathcal{M}^{-1}[M_{v_i}(s)](v_i) = \frac{1}{2\pi j}\int_{c-j\infty}^{c+j\infty} v_i^{-s} M_{v_i}(s)\mathrm{d}s. \tag{4.4}$$

The function $M_{v_i}(s)$ is the Mellin-transformed[1] PDF of the random variable V that has a closed-form Mellin transform expression, $f_V(v_i)$. It is generally holomorphic in the strip $a < \mathrm{Re}(s) < b$. The variable c in (4.4) is a fixed number between a and b that forms the integral path within the strip [13]. If the Bromwich integral conditions [17] are satisfied, then c does not affect the results.

Expression (4.2) is also known as a multiplicative convolution of distributions because its effect is similar to performing a Fourier transform on a linear convolution [18]. Historically, there was a limitation in applying (4.2), because the original Mellin transform was only applicable for positive random variables. However, this problem was resolved by [19], which introduced techniques that can be used to transform random variables in any location on the line in the real domain. This extends the applicability of the Mellin transform to any probability distribution.

The PDF function with realisation v can be divided into two parts such that $f_V(v) = f_V^+(v) + f_V^{-v}$, where $f_V^+(v) = f_V(v)\mathcal{H}(v)$, $f_V^{-v} = f_V(v)\mathcal{H}(-v)$, and $\mathcal{H}(\cdot)$ represents the Heaviside function [18][2]. The Mellin transforms of these functions are shown in equations (4.5) and (4.6).

$$M_{V_i}^+(s) = \mathcal{M}\left[f_{V_i}^+(v)\right](s) \tag{4.5}$$

$$M_{V_i}^-(s) = \mathcal{M}\left[f_{V_i}^-(-v)\right](s) \tag{4.6}$$

Since the PDF $f_{V_i}(v_i)$ of V_i is known, $M_{V_i}^+(s)$ and $M_{V_i}^-(s)$ can be calculated. Both terms can then be used to obtain the Mellin transform of the output Y from equation (4.7) [19].

$$M_Y(s) = \prod_{i=1}^{N_r}\left[M_{V_i}^+(s) + M_{V_i}^-(s)\right] \tag{4.7}$$

[1] The notation $\mathcal{M}$ refers to the Mellin transform equation, while M refers to the Mellin-transformed function.

[2] Readers should take note not to confuse the Heaviside function, denoted by $\mathcal{H}(\cdot)$, with Fox's H-function, denoted as $H(\cdot)$.

The next step is to expand the finite *product of sums* to obtain the finite *sum of products*. The product terms can be collected into two groups: those with a positive parity and those with a negative parity. Positive parity terms are produced by the product of even numbers of $M_{V_i}^{+}(s)$ and even numbers of $M_{V_i}^{-}(s)$. Those terms that do not satisfy the condition of positive parity are classed as having negative parity. By applying the inverse Mellin transform to the positive and negative parity sums, respectively, $f_Y^{+}(y)$ and f_Y^{-y} are obtained, where y is the realisation of the random output variable Y. Finally, the PDF of Y is obtained by adding $f_Y^{+}(y)$ and f_Y^{-y}.

To keep the following discussions in this section succinct, only $f_Y^{+}(y)$ is discussed below. The same procedure is equally valid for f_Y^{-y}. All the transformed PDFs listed in [13], which covers most textbook distributions, can be expressed as products and quotients of gamma functions, denoted by $\Gamma(\cdot)$ in the form:

$$M(s) = \mathcal{C}\eta^{-s}\frac{\prod_{i=1}^{m}\Gamma(b_i + \mathcal{B}_i s)\prod_{i=1}^{n}\Gamma(1 - a_i - \mathcal{A}_i s)}{\prod_{i=n+1}^{p}\Gamma(a_i + \mathcal{A}_i s)\prod_{i=m+1}^{q}\Gamma(1 - b_i - \mathcal{B}_i s)}, \tag{4.8}$$

where $\mathcal{C}$ and η are the appropriate constants to be determined.

A remarkable property of equation (4.8) is that it is a closed form under multiplication and division [14]. This means that after taking a product of two Mellin-transformed PDFs as shown in equation (4.2), the result still has the same form as equation (4.8). In other words, taking the product of any combination of distributions does not change the form of the final result. Another important observation is that (4.8) is the exact form of the Mellin transform of Fox's H-function, denoted by $H(\cdot)$ [14]. In other words, $f_Y^{+}(y)$ can be expressed using the H-function as shown in equation (4.9). Equation (4.10) shows a more succinct notation for the H-function that is commonly used.

$$f_Y^{+}(y) = \frac{\mathcal{C}}{2\pi j}\int_{c-j\infty}^{c+j\infty}[\eta y]^{-s}\frac{\prod_{i=1}^{m}\Gamma(b_i + \mathcal{B}_i s)\prod_{i=1}^{n}\Gamma(1 - a_i - \mathcal{A}_i s)}{\prod_{i=n+1}^{p}\Gamma(a_i + \mathcal{A}_i s)\prod_{i=m+1}^{q}\Gamma(1 - b_i - \mathcal{B}_i s)}\mathrm{d}s, \tag{4.9}$$

$$f_Y^{+}(y) = \mathcal{C}H_{pq}^{mn}\left(\eta y\left|\begin{matrix}(a_1, \mathcal{A}_1), \cdots, (a_p, \mathcal{A}_p)\\(b_1, \mathcal{B}_1), \cdots, (b_q, \mathcal{B}_q)\end{matrix}\right.\right). \tag{4.10}$$

Depending on the values of the parameters, the integral representation of the H-function in (4.9) can be greatly simplified in most situations by using the *Cauchy's residue theorem* [20]. However, the application of the theorem can be very complex. As a result, the evaluation of the H-function should be performed using software with symbolic mathematic capabilities such as Mathematica by Wolfram [21] or

MATLAB by MathWorks [22]. This can be a time-consuming task for design engineers, and therefore it has had limited usage within the professional community.

4.1.2 Applying the Mellin transform to analytical moments of multivariate polynomials

In 2014, the authors of [15] used the Mellin transform method on the moments of multivariate polynomial distributions instead of transforming the PDFs directly.

The known moments for multivariate monomials can be applied to evaluate the moments of multivariate polynomials. The corresponding operation used to deduce the moments of a monomial in equation (4.11) from the knowledge of the higher moments of independent input variables $X_i^{m_i}$ can easily be inferred to be equation (4.12) following the definition and multiplicative convolution of independent variables [23].

$$Y = \prod_i X_i^{m_i} \tag{4.11}$$

$$\mathrm{E}[Y] = \prod_i \mathrm{E}[X_i^{m_i}] \tag{4.12}$$

The case of a polynomial is more tedious, but it follows the same line of reasoning. A multivariate polynomial P can be expressed as equation (4.13), where the set of its constituent multivariate monomials is shown in equation (4.14).

$$P(\boldsymbol{X}) = \sum_{j=0}^{N_r} a_j G_j(\boldsymbol{X}) \tag{4.13}$$

$$G_j(\boldsymbol{X}) = \prod_{i \in I_j} X_i^{m_i} \tag{4.14}$$

The kth power of P, denoted by P^k, can then be expressed as:

$$P^k(\boldsymbol{X}) = \sum_{j=0}^{kN_r} a_j G_j(\boldsymbol{X}) \tag{4.15}$$

for some appropriate value a_j and monomials G_j. The indexing variables i and j as well as the values of m_i and a_j are not important in this discussion, and therefore they are kept in an abstract form here.

By exploiting the linearity of the expectation operator and multiplicative convolution of independent variables [23], the kth-order moment of the multivariate polynomial P can therefore be further simplified to:

$$\mathrm{E}[P^k] = \sum_{j=0}^{kN_r} a_j \left(\prod_{i \in I_j} \mathrm{E}[X_i^{m_i}] \right). \tag{4.16}$$

Equation (4.16) is significant because it allows the calculation of $E[P^k]$ through purely algebraic operations via the expectation of the higher-order moments to the mth order of the independent input variables $X_i^{m_i}$.

Rewriting equations (4.5) and (4.6) gives:

$$M_V^+(s) = \mathcal{M}\left[f_V^+(v)\right](s)$$

$$= \int_0^\infty v^{s-1} f_V^+(v)\mathrm{d}v \tag{4.17}$$

and

$$M_V^-(s) = M\left[f_V^-(-v)\right](s)$$

$$= \int_0^\infty v^{s-1} f_V^-(-v)\mathrm{d}v \cdot \tag{4.18}$$

Assume that $z = v^m$ such that z is a subset of real values $z \in \mathbb{R}$, and m is a subset of a set of integers $m \in \mathbb{Z}$. The PDF of z can be found directly through the change of variable technique. There are two cases to consider.

1. When m is an odd integer, $z = v^m$ is a one-to-one and monotonic mapping; therefore, $f_Z^+(z)$ and $f_Z^-(z)$ can be rewritten as equations (4.19) and (4.20).

$$f_Z^+(z) = \left|\frac{\mathrm{d}v}{\mathrm{d}z}\right| f_V^+(v)\Big|_{v=z^{\frac{1}{m}}}$$

$$= \frac{1}{m} z^{\frac{1-m}{m}} f_V^+\left(z^{\frac{1}{m}}\right) \tag{4.19}$$

$$f_Z^-(z) = \left|\frac{\mathrm{d}v}{\mathrm{d}z}\right| f_V^-(v)\Big|_{v=z^{\frac{1}{m}}}$$

$$= \frac{1}{m} z^{\frac{1-m}{m}} f_V^-\left(z^{\frac{1}{m}}\right). \tag{4.20}$$

2. When m is an even integer, $z = v^m$ is a two-to-one mapping $z = v^m = (-v)^m$, and therefore $f_Z^+(z)$ and $f_Z^-(z)$ can be rewritten as equations (4.21) and (4.22).

$$f_Z^+(z) = \left|\frac{\mathrm{d}v}{\mathrm{d}z}\right| f_V^+(v)\Big|_{v=z^{\frac{1}{m}}} + \left|\frac{\mathrm{d}v}{\mathrm{d}z}\right| f_V^-(v)\Big|_{v=-z^{\frac{1}{m}}}$$

$$= \frac{1}{m} z^{\frac{1-m}{m}}\left(f_V^+\left(z^{\frac{1}{m}}\right) + f_V^-\left(-z^{\frac{1}{m}}\right)\right) \tag{4.21}$$

$$f_Z^-(z) = 0. \tag{4.22}$$

Generalising, the expectation of $z = v^m$ is:

$$\mathrm{E}[v^p] = \int_0^\infty z f_Z^+(z)dz + \int_{-\infty}^0 z f_Z^-(z)dz$$

$$= \int_0^\infty v^m f_V^+(v)dv + (-1)^m \int_0^\infty v^m f_V^-(-v)dv$$

$$= M_V^+(m+1) + (-1)^m M_V^-(m+1). \tag{4.23}$$

This allows $\mathrm{E}[v^p]$ to be defined separately for each type of distribution. For example,

1. *Symmetrical distributions:* due to symmetry, $f_V^+(v) = f_V^{-v}$, so that $M_V^+(\mathrm{s}) = M_V^-(\mathrm{s})$. If m is odd, $\mathrm{E}[V^m] = 0$, and if m is even, $\mathrm{E}[V^m] = 2M_V^+(m+1)$.
2. *One-sided distributions*: the half-distribution is multiplied by two to ensure that the area under the PDF is unity. If $f_V^-(v) = 0$ (positive support) $\Rightarrow M_V^-(s) = 0$, $\mathrm{E}[V^m] = 2M_V^+(m+1)$. And if $f_V^+(v) = 0$ (negative support) $\Rightarrow M_V^+(s) = 0$, $\mathrm{E}[V^m] = (-1)^m 2M_V^-(m+1)$.

Therefore, by making use of equation (4.23), $\mathrm{E}[X^m]$ can be predefined for m under different constraints through the creation of a so-called 'Mellin transform lookup table' containing the Mellin-transformed PDFs of standard distributions from [13]. This table is given in appendix B.

To replace the complex evaluation of the Fox's H-function, equation (4.10) is rewritten so that the general-form PDF of a random variable V which follows a standard tabulated distribution with realisation v can be represented by:

$$f_V(v) = CH_{pq}^{mn}\left(\eta v \left| \begin{matrix} (a_1, \mathcal{A}_1), \cdots, (a_p, \mathcal{A}_p) \\ (b_1, \mathcal{B}_1), \cdots, (b_q, \mathcal{B}_q) \end{matrix} \right.\right), \tag{4.24}$$

where C and η are the corresponding constants and the other notations are kept in abstract form.

Using the ith-order raw moment $\mathrm{E}[X^i]$ of a random variable X from equation (3.19) discussed in the previous chapter, the rth-order moment of $f_V(v)$ can be written as:

$$\int v^r f_V(v)dv = \frac{C\eta^{-(r+1)} \prod_{i=1}^{m} \Gamma(b_i + \mathcal{B}_i(r+1)) \prod_{i=1}^{n} \Gamma(1 - a_i - \mathcal{A}_i(r+1))}{\prod_{i=n+1}^{p} \Gamma(a_i + \mathcal{A}_i(r+1)) \prod_{i=m+1}^{q} \Gamma(1 - b_i - \mathcal{B}_i(r+1))}. \tag{4.25}$$

Equation (4.25) has the same form of expression as those of the Mellin-transformed PDFs given in appendix B. This similarity is not a coincidence, since the definition of the Mellin transform in equation (4.3) is almost identical to the definition of moments, except for the range of integration. Therefore, when the Mellin transform function has the form of equation (4.25), this means that the rth moment of V with PDF $f_V(v)$ can be determined by substituting $s = r + 1$ into equation (4.8). This is true even for the sum and/or products of multiple random variables $\mathbf{V} = \{V_1, \ldots, V_{N_r}\}$ that follow the standard tabulated distributions.

In other words, if Y is expressed as a multivariate polynomial of the form shown in equation (4.13), its high-order analytical moments can be obtained without evaluating the PDF explicitly. Instead, it suffices to look up the corresponding expression in appendix B. The next two subsections relate the random variables X and V to support the appropriate use of the Mellin transform lookup table.

4.1.3 Moment calculation for positive-order variables, $m \in \mathbb{Z}^+$

The preceding subsections showed that the mth-order moment of X, $\mathrm{E}[X^m]$, is an important intermediate quantity in the calculation used to determine the high-order analytical moments of a polynomial. It has also been established that $\mathrm{E}[X] = \mathrm{E}[V]$ can be deduced from $\mathcal{M}[f_X](s)$ by substituting a suitable value for s with $m + 1$.

The following discussions outline the process of calculating $\mathrm{E}[X^m]$. Note that the term 'standard distributions' refers to the distributions listed in appendix B, where the direct Mellin transform has been predefined. Random variables distributed according to standard distributions are denoted by V and their realisation by v.

4.1.3.1 Standard distributions

$\mathrm{E}[V^m]$ cannot be evaluated by a direct application of multiplicative convolution, because V is fully correlated to itself, thus violating the independence assumption. Hence, $\mathrm{E}[V]^m \neq \mathrm{E}[V^m]$. By applying equation (4.23), $\mathrm{E}[V^m]$, $v \in \mathbb{R}$ and $m \in \mathbb{Z}^+$ can be inferred given the knowledge of the Mellin transform $\mathcal{M}[V]$, where $\mathbb{R}$ and $\mathbb{Z}^+$ denote the set of real numbers and positive integers, respectively.

Almost all standard distributions listed [13, 16, 24–26] belong to either symmetrical or one-sided distributions. Therefore, the simplified expressions for these special cases are included in table 4.1. All odd-order moments of symmetrical distributions vanish, resulting in significant simplification.

Table 4.1. The expectation $E[V^m]$ for $v \in \mathbb{R}$ and $m \in \mathbb{Z}^+$ where $\mathbb{R}$ and $\mathbb{Z}^+$ denote the set of real numbers and positive integers, respectively.

Distribution type		$\mathrm{E}[V^m]$
Any distribution		$M_V^+(m+1) + (-1)^m M_V^-(m+1)$
Symmetrical distribution	m is odd	0
	m is even	$2M_V^+(m+1)$
One-sided distribution	Positive support	$2M_V^+(m+1)$
	Negative support	$(-1)^m 2M_V^-(m+1)$

4.1.3.2 Translated standard distributions

When the Mellin transform is applied to the distribution of realistic random variables such as a translated normal distribution, its limitation becomes apparent. The Mellin transform of the central normal distribution shown in appendix B is invalid if the distribution is translated.

In general, any measurement can be modelled by $X = \tau + V$, where τ represents the translation value of the measurement while V represents a random variable that follows the standard distribution to describe the uncertainty of X. Note that τ is equivalent to the mean if the underlying distribution is symmetrical. Using binomial expansion:

$$X^m = \sum_{p=0}^{m} \binom{m}{p} \tau^{m-p} V^p \tag{4.26}$$

for $m > 1$ gives an expression for $\mathrm{E}[X^m]$, as follows:

$$\mathrm{E}[X^m] = \sum_{p=0}^{m} \binom{m}{p} \tau^{m-p} \mathrm{E}[V^p]. \tag{4.27}$$

The convention $\mathrm{E}[V^0] = 1$ is adopted here. Thus, $\mathrm{E}[X^m]$ can be found by algebraic manipulation because $\mathrm{E}[V^p]$ can be calculated from appendix B and table 4.1 for any degree $m \in \mathbb{Z}^+$.

4.1.4 Moment calculation for negative-order variables, $m \notin \mathbb{Z}^+$

The constraint $m \in \mathbb{Z}^+$ in the previous subsection covers the class of all polynomial functions. This subsection considers the extension of $m \notin \mathbb{Z}^+$, where the evaluation of high-order moments can be done analytically. This is possible because the initial formulation of the Mellin transform only requires that m be a real number. Therefore, the use of the proposed framework can be extended to a more general class of functions than multivariate polynomials.

4.1.4.1 Standard distributions

Negative-order variables restrict the set of distributions for which the moment can be calculated analytically, because equation (4.26) cannot be applied for a translated distribution when $m \notin \mathbb{Z}^+$. Thus, only the moments from standard distributions can be analytically determined when $m \notin \mathbb{Z}^+$. The calculation of $\mathrm{E}[X^m]$ follows directly from the evaluation of $\mathrm{E}[V^m]$, which is obtained from the one-sided distribution in table 4.1. Table 4.2 shows admissible combinations of m and X, excluding cases in which the evaluation of X^m would result in a complex number or division by zero.

4.1.4.2 Exceptions for distributions with compact support

The previous subsection stated that the evaluation of $\mathrm{E}[X^m]$ for $m \notin \mathbb{Z}^+$ is only possible for random variables with standard distributions. However, there are

Table 4.2. The expectation $E[X^m]$ for a wider class of functions for $m \notin \mathbb{Z}^+$, where $\mathbb{Z}^+$ denotes the set of positive integers.

Set of m	Admissible X	$E[X^m]$
Negative integer, $m \in \mathbb{Z}^-$	$X > 0$	$2M_V^+(m+1)$
	$X < 0$	$(-1)^m 2M_V^-(m+1)$
Non-integer real number, $m \in \mathbb{R}\mathbb{Z}$	$X > 0$	$2M_V^+(m+1)$

exceptions to this limitation. Specifically, if either $X > 0$ or $X < 0$, then the Mellin transform for random variables with compact support distributions can be expressed in closed form for any real value m, even when it is a translated distribution. This is a very important exception because compact support distributions represent some of the most commonly used distributions in uncertainty evaluation, such as uniform, beta, Weibull, or lognormal distributions, etc.

In short, provided that $X > 0$ or $X < 0$, $E[X^m]$ for $m \notin \mathbb{Z}^+$ for uniform, beta, Weibull, and lognormal distributions can be derived from equation (3.19).

4.1.4.3 No exceptions for distributions with global support

Strictly speaking, $E[X^m]$ for $m \notin \mathbb{Z}^+$ where $(m < -1)$ cannot be modelled by a normal distribution (or any other distributions that have global support) because various moments of the output function Y would be undefined, leading to hidden logical contradictions within the analysis. The mathematical proof is given below.

Let the variable X be normally distributed. The higher-order moments of X^{-1} can be expressed as $E[X^{-m}] = \int_{-\infty}^{\infty} x^{-m} f_X(x)\mathrm{d}x$, where $f_X(x)$ represents the PDF of X. Without loss of generality, we assume that $X \sim \frac{1}{\sigma\sqrt{2\pi}}\exp\left(-\frac{(x-\mu)^2}{2\sigma^2}\right)$, where $\mu > 0$, $\sigma > 0$, and $m > 0$. The proofs for odd and even m are slightly different.

In the case where m is even, x^{-m} is positive definite,

$$E[X^{-m}] = \frac{1}{\sigma\sqrt{2\pi}}\int_{-\infty}^{\infty} x^{-m}\exp\left(-\frac{(x-\mu)^2}{2\sigma^2}\right)\mathrm{d}x$$

$$= \frac{1}{\sigma\sqrt{2\pi}}\int_{-\infty}^{0} x^{-m}\exp\left(-\frac{(x-\mu)^2}{2\sigma^2}\right)\mathrm{d}x + \frac{1}{\sigma\sqrt{2\pi}}\int_{0}^{\infty} x^{-m}\exp\left(-\frac{(x-\mu)^2}{2\sigma^2}\right)\mathrm{d}x$$

$$\geqslant \frac{1}{\sigma\sqrt{2\pi}}\left\{\int_{-\infty}^{0} x^{-m}\exp\left(-\frac{(x-\mu)^2}{2\sigma^2}\right)\mathrm{d}x + \exp\left(-\frac{\mu^2}{2\sigma^2}\right)\int_{0}^{2\mu} x^{-m}\mathrm{d}x\right.$$

$$\left. + \int_{2\mu}^{\infty} x^{-m}\exp\left(-\frac{(x-\mu)^2}{2\sigma^2}\right)\mathrm{d}x\right\}$$

$$\geqslant \frac{\exp\left(-\frac{\mu^2}{2\sigma^2}\right)}{\sigma\sqrt{2\pi}} \int_0^{2\mu} x^{-m} \mathrm{d}x$$

$$= \frac{\exp\left(-\frac{\mu^2}{2\sigma^2}\right)}{\sigma\sqrt{2\pi}(-m+1)} x^{-m+1} \Big|_0^{2\mu}. \tag{4.28}$$

The function $x^{-m+1} \Big|_0^{2\mu}$ does not converge to any finite value for any even number $m > 1$, hence $\mathrm{E}[X^{-m}]$ does not exist.

In the case where m is odd,

$$\mathrm{E}[X^{-m}] = \frac{1}{\sigma\sqrt{2\pi}} \int_{-\infty}^{\infty} x^{-m} \exp\left(-\frac{(x-\mu)^2}{2\sigma^2}\right) \mathrm{d}x$$

$$= \frac{1}{\sigma\sqrt{2\pi}} \int_{-\infty}^{0} x^{-m} \exp\left(-\frac{(x-\mu)^2}{2\sigma^2}\right) \mathrm{d}x + \frac{1}{\sigma\sqrt{2\pi}} \int_{0}^{\infty} x^{-m} \exp\left(-\frac{(x-\mu)^2}{2\sigma^2}\right) \mathrm{d}x$$

$$= \frac{1}{\sigma\sqrt{2\pi}} \int_{0}^{\infty} x^{-m} \left[\exp\left(-\frac{(x-\mu)^2}{2\sigma^2}\right) - \exp\left(-\frac{(x+\mu)^2}{2\sigma^2}\right)\right] \mathrm{d}x$$

$$= \frac{2\exp\left(-\frac{\mu^2}{2\sigma^2}\right)}{\sigma\sqrt{2\pi}} \int_{0}^{\infty} x^{-m} \exp\left(-\frac{x^2}{2\sigma^2}\right) \sinh\left(\frac{\mu x}{\sigma^2}\right) \mathrm{d}x, \tag{4.29}$$

$\mathrm{E}[X^{-m}]$ converges only when $m < 2$. In other words, the integral does not converge for all odd $m > 2$. A similar argument applies to other global support distributions, such as Student's t distribution and the Laplace distribution.

4.1.5 Extension to high-dimensional correlated variables

The previous subsections assume that the random variables are uncorrelated. In cases where correlated variables are involved, eigentransformation can be used to map these dependent variables into an equivalent set of independent variables, the underlying mathematics of which are identical to those of principal component analysis [27].

For engineering applications, a covariance matrix is commonly used to define the correlation between variables because there is typically insufficient information to explicitly establish the joint PDF. This is especially true in the case of high-dimensional correlation.

The eigenvectors of the covariance matrix are used to form a linear transformation matrix $\mathbf{A}$ of size $N_r \times N_r$ [28]. By applying the orthogonal transformation, equation (4.30) forms a set of statistically uncorrelated random variables $\mathbf{Z} = \{Z_1, \ldots, Z_{N_r}\}$ in which all random variables $\mathbf{Z}$ are normally distributed [28, 29].

$$\mathbf{Z} = \mathbf{AX} = \begin{bmatrix} a_{11} & \cdots & a_{1N_r} \\ \vdots & \ddots & \vdots \\ a_{N_r1} & \cdots & a_{N_rN_r} \end{bmatrix} \mathbf{X} = \begin{bmatrix} a_{11}X_1 + \ldots + a_{1N_r}X_{N_r} \\ \vdots \\ a_{N_r1}X_1 + \ldots + a_{N_rN_r}X_{N_r} \end{bmatrix}, \tag{4.30}$$

The joint PDF can also be approximated using a Gaussian copula [29], after which the Nataf transformation [30] is applied to convert the correlated input variables **X** into correlated standard normal variables. Once this has been achieved, the standard linear transformation is performed to obtain uncorrelated standard normal distributions **Z**, as elaborately described in [29]. This method results in the transformation matrix **A** as the lower triangular matrix from Cholesky factorisation [31] of the covariance matrix of correlated normal distributions.

Both of these approaches have been reported to produce the same results [29]. Employing either of these methods allows high-dimensional correlated variables to be expressed as uncorrelated standard normal distributions, which in turn facilitates the applicability of the moment evaluation described in the previous subsections.

4.2 A toolbox for moment-based standard uncertainty evaluation

To streamline the process of computing standard uncertainty using the methods described in section 4.1, this subsection introduces a user-friendly toolbox called the *Analytical Uncertainty Calculator* (ANUNCEATOR), which is now freely available at http://polymoment.com/Analytical_Uncertainty_Calculator. It uses Python's symbolic mathematical library [32] to enable rapid and reliable calculation of the variance. The toolbox simplifies the process of calculating the output variance through three simple steps.

First, a user starts by selecting a number of inputs to their system using the 'Add' button. The user can adjust the statistical properties, i.e. the distribution and symmetry type, of every input by clicking on it (see figure 4.2).

After all the inputs have been added, the system model is added into the text box. Figure 4.3 shows this step for an example system:

$$Y = X_1^2 + X_2^2, \tag{4.31}$$

where both the random inputs follow the normal distribution within their respective standard deviations (scale parameters) of $s1$ and $s2$ with their respective means (location parameter) of $m1$ and $m2$.

The variance $u^2[Y]$ is evaluated and simplified (as shown in figure 4.4) as:

$$u^2[Y] = 4m1^2s1^2 + 4m2^2s2^2 + 2s1^4 + 2s2^4. \tag{4.32}$$

Upon calculation, the result output from the toolbox can be copied into a file (using the copy–paste function) for future reference. The toolbox also features additional functions to evaluate the skewness and kurtosis of the system by clicking on the 'Skewness' and 'Kurtosis' buttons, respectively, as shown in figures 4.5(a) and (b). Alternatively, users can compute moments up to the 10th raw moment of the output Y by selecting the moment order and clicking on the 'Moment', as illustrated in figure 4.5(c).

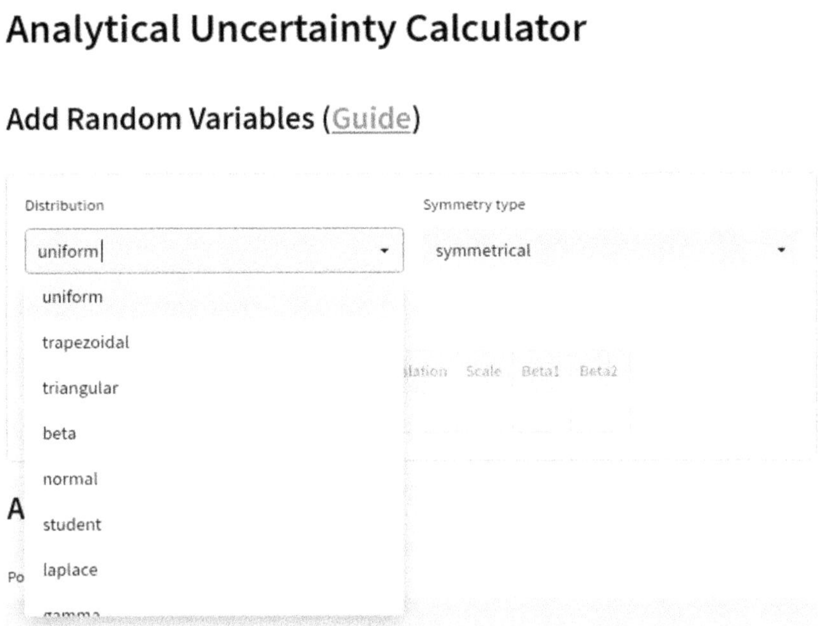

Figure 4.2. Adjusting the statistical properties (the distribution and symmetry type) of a variable in the ANUNCEATOR user interface.

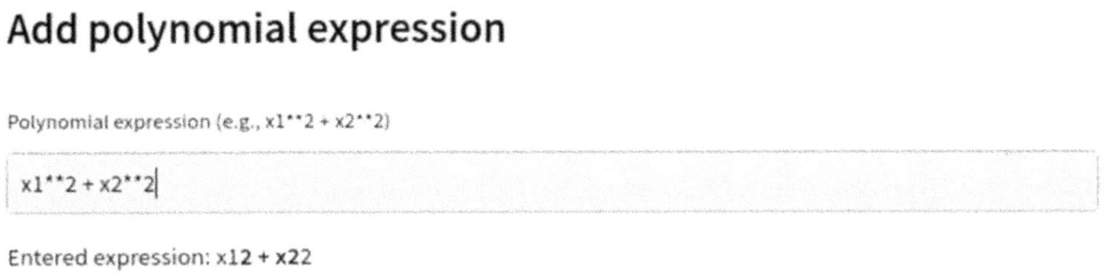

Figure 4.3. A system modelled as a polynomial $Y = X_1^2 + X_2^2$ is added into the ANUNCEATOR toolbox.

4.3 Case studies for moment-based analytical standard uncertainty evaluation

4.3.1 Case study 1—monomial: magnetic force microscope

A magnetic force microscope reports the magnetic properties of a given material by measuring the change in a cantilever's oscillation frequency via the measurement function shown in equation (4.33).

$$\Delta f = \frac{\pi M d^3}{6c}, \tag{4.33}$$

where Δf is the frequency shift, M is the saturation magnetisation, d is the diameter of the nanoparticles, and c is a tip calibration factor adjusted for different heights of

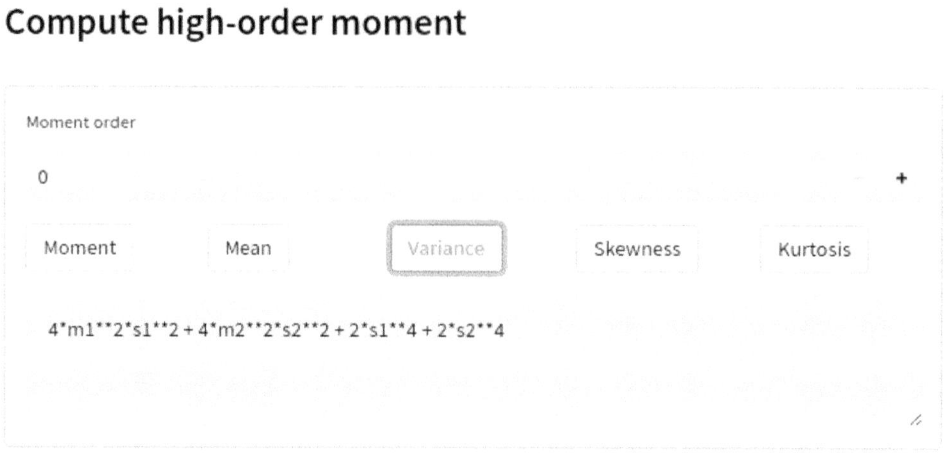

Figure 4.4. ANUNCEATOR displaying the variance of system $Y = X_1^2 + X_2^2$.

the tip from the sample. The variable c^{-1} belongs to the class of functions described in section 4.1.4, where $m \notin \mathbb{Z}^+$.

The authors of [33] calibrated this microscope using the set of expected values and ranges of each variable. However, the paper did not specify any distribution. Hence, the distributions that fit the given descriptions are assumed and listed in table 4.3, where c is assumed to be calibrated at a lift height of 50 nm. Since μ_d has been fixed, the shape parameter of the gamma distribution is calculated from $\beta = U[d]\mu_d/\sigma_d$.

Using the GUM linear approximation method given in section 2.2 [34], the standard uncertainty U_G is computed as shown in equation (4.34), with G denoting the GUM method.

$$U_G[\Delta f] = \frac{\pi}{6}\sqrt{\left(\frac{\mu_d^3}{\mu_c}\right)^2 U^2[M] + \left(\frac{3\mu_M\mu_d^2}{\mu_c}\right)^2 U^2[d] + \left(-\frac{\mu_M\mu_d^3}{\mu_c^2}\right)^2 U^2[c]}$$

$$= \frac{\pi}{6}\sqrt{\left(\frac{\mu_d^3}{\mu_c}\right)^2 \frac{\sigma_M^2}{3} + \left(\frac{3\mu_M\mu_d^2}{\mu_c}\right)^2 \mu_d\sigma_d + \left(-\frac{\mu_M\mu_d^3}{\mu_c^2}\right)^2 \frac{\sigma_c^2}{3}}. \tag{4.34}$$

Next, using the procedure outlined in figure 4.1, the standard uncertainty U_A is computed using the analytical moment-based method. First, $(\Delta f)^2$ is determined in terms of the inputs by expanding from equation (4.33):

$$(\Delta f)^2 = \frac{\pi^2 M^2 d^6}{36c^2}. \tag{4.35}$$

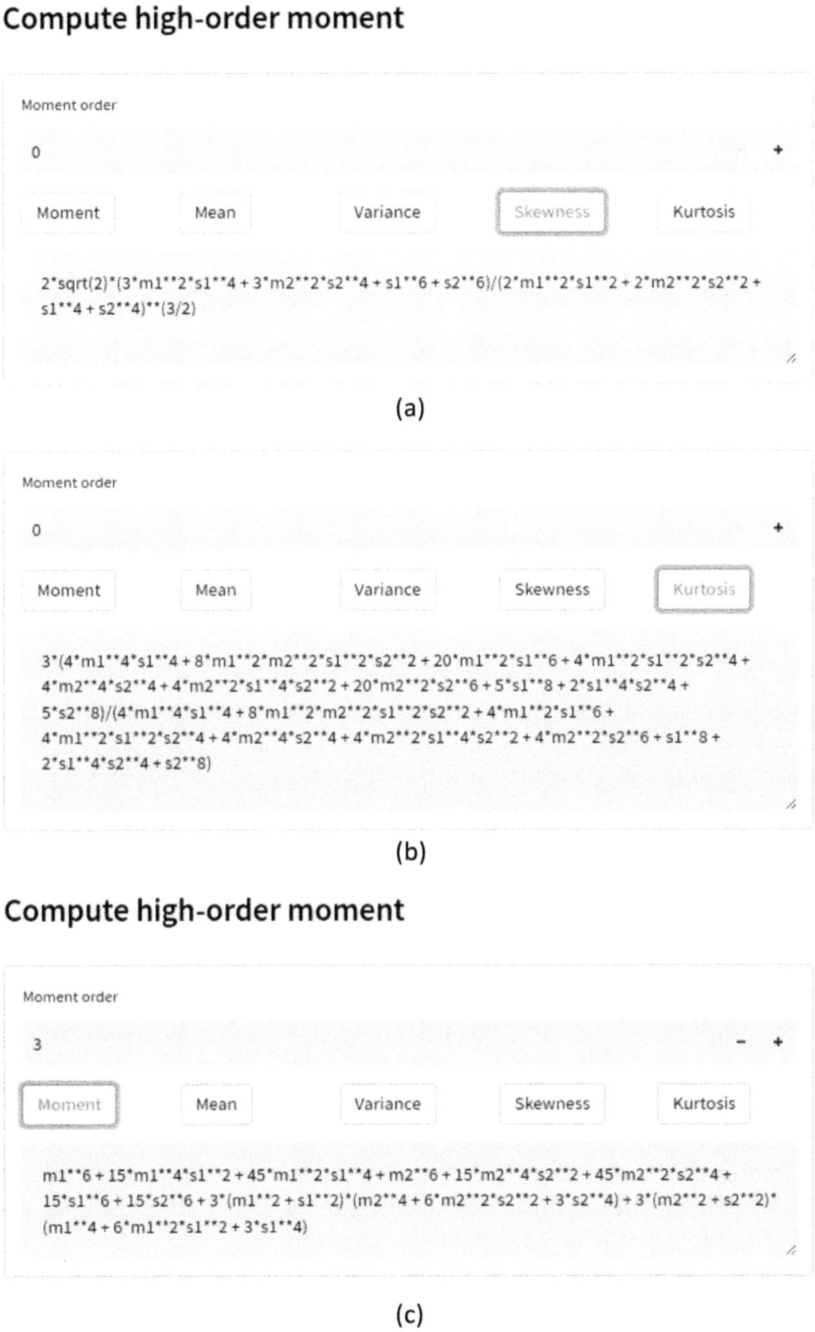

Figure 4.5. ANUNCEATOR displaying the output for system $Y = X_1^2 + X_2^2$, including: (a) skewness; (b) kurtosis; and (c) third-order moment.

Table 4.3. Variables for the magnetic force microscope example.

Variable, X	PDF	Expected value (μ)	$U[X]$	Range
M	Uniform	25×10^{-5} A nm^{-1}	$\sqrt{\frac{\sigma_M^2}{3}}$	$0 < \sigma_M < 1 \times 10^{-5}$
d	Gamma, positive only	17 nm	$\sqrt{\mu_d \sigma_d}$	$1 < \sigma_d < 3$
c	Uniform	1.14 A nm^2 Hz^{-1}	$\sqrt{\frac{\sigma_c^2}{3}}$	$0.01 < \sigma_c < 0.33$

The standard uncertainty $U_A[\Delta f]$ can be obtained by substituting equation (4.35) into equation (2.8) from section 2.5:

$$U_A[\Delta f] = \sqrt{E[(\Delta f)^2] - E^2[\Delta f]}$$

$$= \frac{\pi}{6}\sqrt{E[M^2]E[d^6]\,E[c^{-2}] - (E[M]E[d^3]E[c^{-1}])^2}\,. \tag{4.36}$$

Equation (4.36) allows the standard uncertainty $U_A[\Delta f]$ to be directly computed from the terms $E[M^2]$, $E[d^6]$, $E[c^{-2}]$, $E[M]$, $E[d^3]$, and $E[c^{-1}]$. Given that all these terms have the form $E[X^m]$, the procedure in figure 4.1 can easily be followed.

1. The input variable M is uniformly distributed (table 4.3) with positive-order variables, $m \in \mathbb{Z}^+$. Thus equation (4.27) and *row 2* of table 4.1 are applied.
2. The input variable d is a one-sided gamma distribution with positive-order variables, $m \in \mathbb{Z}^+$, thus equation (4.27) and *row 3* of table 4.1 are applied.
3. The input variables c^{-1} and c^{-2} have negative-order variables, $m \notin \mathbb{Z}^+$. Both are uniformly distributed; therefore, they require the use of table 4.2.

Using the Mellin transform lookup table in appendix B, the appropriate $F_V^+(m+1)$ are then obtained and substituted into equation (4.36) to obtain equation (4.37):

$$U_A[\Delta f] = \frac{\pi}{6}\sqrt{\left[\mu_M^2 + \frac{\sigma_M^2}{3}\right]\left[\frac{\sigma_d^6\left(\frac{\mu_d}{\sigma_d}+5\right)_6}{\mu_c^2 - \sigma_c^2}\right] - \frac{\mu_M^2\sigma_d^6}{4\sigma_c^2}\left[\left(\frac{\mu_d}{\sigma_d}+2\right)_3\right]^2 \log^2\left(\frac{\mu_c+\sigma_c}{\mu_c-\sigma_c}\right)}\,. \tag{4.37}$$

These procedures for obtaining $U_A[\Delta f]$ can be achieved using the toolbox described in section 4.2. Thus, although the derivation appears to be more tedious compared to that of $U_G[\Delta f]$, it is not more difficult to accomplish.

The variances of the inputs, σ_d, σ_c and σ_M can be independently varied for a more thorough analysis. Figure 4.6(a) shows an example of a one-dimensional sampling curve in σ_d–σ_c–σ_M space in which the sampled points are indexed from one to 20, forming the horizontal axis of figure 4.6(b). Figure 4.6(b) shows the standard uncertainties obtained via the GUM linear approximation method $U_G[\Delta f]$ and the analytical moment-based standard uncertainty estimation framework $U_A[\Delta f]$ benchmarked against the MC method.

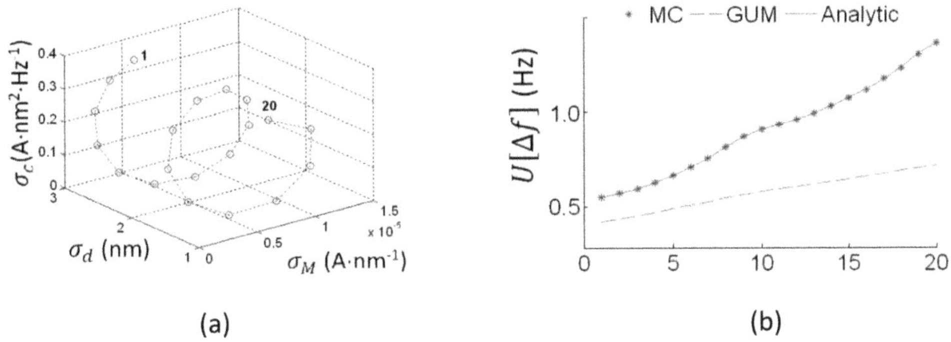

Figure 4.6. (a) Test points 1–20 in the σ_d–σ_c–σ_M space. (b) Comparing standard uncertainty obtained through: *MC*—Monte Carlo, $U_G[\Delta f]$—*GUM* and $U_A[\Delta f]$—*analytic* methods.

The uncertainty of the measured frequency shift Δf of the magnetic force microscope is consistently underestimated. Accordingly, the GUM [34] advocates that the MC simulation method from section 2.3 should be applied instead. However, the advantage of using the analytical moment-based method as outlined in figure 4.1 is that the exact analytical expression of standard uncertainty $U_A[\Delta f]$ is obtained, which the MC method does not produce. In this example, equation (4.37) provides an insight into the interplay and relative contributions of different input factors to the final uncertainty of the frequency shift. If sampling is involved in the determination of distribution parameters (Type A evaluation), it can be explicitly included in the final expression of uncertainty, which is especially useful for high-precision scientific measurements.

4.3.2 Case study 2—simple polynomial: microwave meter calibration

The calibration measurement of microwave devices is a well-known example from Supplement 1 of the GUM [35]. It shows the advantage of the MC simulation method over the linear approximation method [34]. The measurement function is:

$$\delta Y = X_1^2 + X_2^2, \tag{4.38}$$

where δY measures the quality of the microwave power meter that is calibrated with respect to the standard microwave power meter. X_1 and X_2 are the real and imaginary components of the reflection coefficient of the power meter to be calibrated. Units of measurement are not required for these variables.

Table 4.4 shows the distributions and parameter values from [35]. The standard uncertainty U_G is computed from equation (4.34) using the GUM linear approximation method:

$$U_G[\delta Y] = \sqrt{2\mu_{X_1}\ U^2[X_1] + 2\mu_{X_2}\ U^2[X_2]}$$

$$= \sqrt{\left[2\mu_{X_1}\sigma_{X_1}\right]^2 + \left[2\mu_{X_2}\sigma_{X_2}\right]^2}. \tag{4.39}$$

Table 4.4. The variables for the microwave meter calibration example [35].

	Variable, X	PDF	Expected value (μ)	Standard uncertainty, $U[X]$	Range
Test 1	X_1	Normal	0.000	σ_{X_1}	$0 < \sigma_{X_1} < 0.05$
	X_2	Normal	0.000	σ_{X_2}	$0 < \sigma_{X_2} < 0.05$
Test 2	X_1	Normal	0.010	σ_{X_1}	$0 < \sigma_{X_1} < 0.05$
	X_2	Normal	0.000	σ_{X_2}	$0 < \sigma_{X_2} < 0.05$
Test 3	X_1	Normal	0.050	σ_{X_1}	$0 < \sigma_{X_1} < 0.05$
	X_2	Normal	0.000	σ_{X_2}	$0 < \sigma_{X_2} < 0.05$

The standard uncertainty U_A for multivariate polynomials is next calculated using figure 4.1, as follows:

$$U_A[\delta Y] = \sqrt{E[(\delta Y)^2] - E^2[\delta Y]}$$

$$= \sqrt{E[X_1^4] + E[X_2^4] - E^2[X_1^2] - E^2[X_2^2]}$$

$$= \sqrt{2\sigma_{X_1}^2\left(2\mu_{X_1}^2 + \sigma_{X_1}^2\right) + 2\sigma_{X_2}^2\left(2\mu_{X_2}^2 + \sigma_{X_2}^2\right)}. \tag{4.40}$$

Figure 4.7 shows the sampling curve and a comparison between the MC, linear approximation, and analytical moment methods, respectively, for the three tests outlined in Supplement 1 of the GUM [35]. In all three tests, $U_G[\delta Y]$ failed to accurately estimate the standard uncertainties, because it does not consider the higher-order terms $\sigma_{X_1}^4$ and $\sigma_{X_2}^4$. This resulted in a significant underestimation of error when σ_{X_i} was comparable to μ_{X_i}, which became increasingly severe as $\mu_{X_i} \to 0$. In contrast, the analytical moment-based method $U_A[\delta Y]$ performed comparably to the MC simulation method and offered the advantage of having an explicit expression for its standard uncertainty.

4.3.3 Case study 3—high-order Taylor series approximation: eddy current measurement

The thickness of a metal coating can be non-destructively determined through the measurement of the voltage that is induced by eddy currents via the measurement function in equation (4.41) [36]:

$$T = -T_0 \log\left(\frac{V}{V_k}\right), \tag{4.41}$$

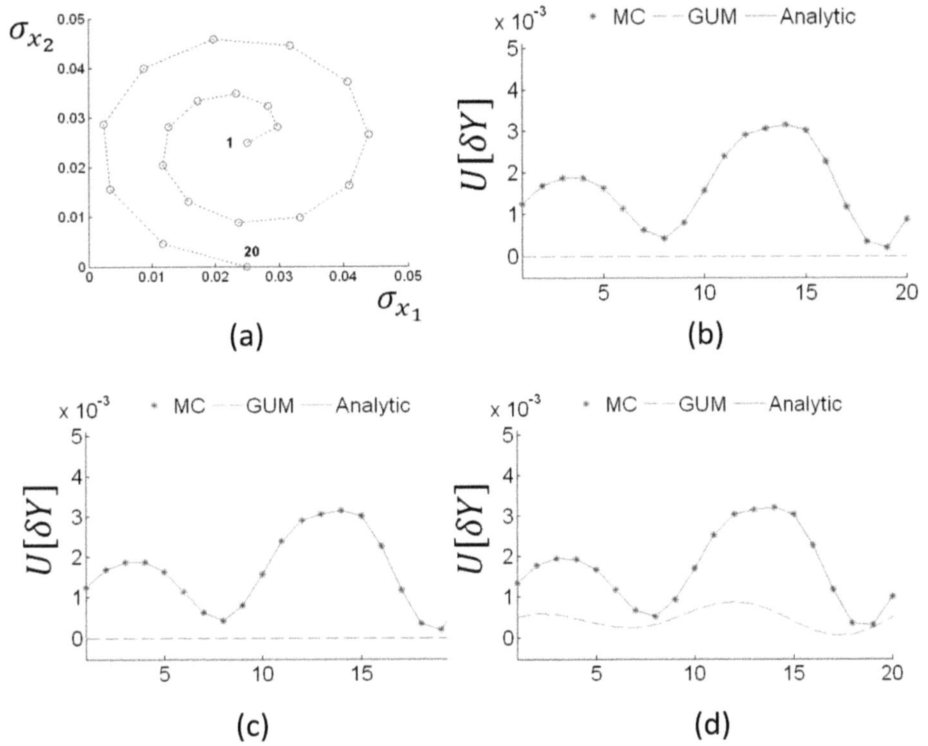

Figure 4.7. (a) Test points 1–20 in σ_1–σ_2 space for the tests from [35]: (b) Test 1, (c) Test 2, and (d) Test 3 obtained using: *MC*—Monte Carlo, $U_G[\Delta f]$—*GUM*, and $U_A[\Delta f]$—*analytic* methods.

where T is the thickness of the metallic layer, V is the measured voltage, and T_0 and V_k are constants for the measurement system to be calibrated during the characterisation phase, such that $V_k > V$.

Equation (4.41) is further simplified to (4.42) by setting the normalisation constant $T_0 = 1$ and introducing a new variable $R = \frac{V}{V_k} - 1$ whose range is $-1 < R < 0$. Table 4.5 shows the three independent test cases at different mean operating points μ_R of the measurements with the standard uncertainty of σ_R set to less than 1%.

$$T = -\log(1 + R) \tag{4.42}$$

As in the previous two case studies, the standard uncertainty U_G for this measurement is shown in equation (4.43), which was obtained using equation (4.34).

$$U_G[T] = \frac{\sigma_R}{1 + \mu_R} \tag{4.43}$$

Since the measurement function is nonlinear, a polynomial approximation is required in order to apply the analytical moment-based method. A simple third-order Taylor series expansion was therefore chosen, as shown in equation (4.44).

Table 4.5. The tests and variables for the eddy current measurement.

Test	Variable, X	PDF	Expected value (μ)	Standard uncertainty, $U[X]$	Range
Test 1	R	Normal	−0.1	σ_R	$0 < \sigma_R < 1 \times 10^{-3}$
Test 2	R	Normal	−0.3	σ_R	$0 < \sigma_R < 1 \times 10^{-3}$
Test 3	R	Normal	−0.5	σ_R	$0 < \sigma_R < 1 \times 10^{-3}$

The analytical moment-based estimation of standard uncertainty for the third-order Taylor approximation is shown in (4.45).

$$T \approx -R + \frac{1}{2}R^2 - \frac{1}{3}R^3 \tag{4.44}$$

$$U_A[T] = \sqrt{E[(T)^2] - E^2[T]}$$

$$= \sqrt{E\left[\left(-R + \frac{1}{2}R^2 - \frac{1}{3}R^3\right)^2\right] - \left[E\left(-R + \frac{1}{2}R^2 - \frac{1}{3}R^3\right)\right]^2}$$

$$= \sqrt{\frac{5}{3}\sigma_R{}^6 + \left(4\mu_R{}^2 - 4\mu_R + \frac{5}{2}\right)\sigma_R{}^4 + \left(\mu_R{}^4 - 2\mu_R{}^3 + 3\mu_R{}^2 - 2\mu_R + 1\right)\sigma_R{}^2} \tag{4.45}$$

Figure 4.8(a) compares the third-order approximation accuracy against the original function, along with the fifth-order Taylor series expansion. Here, it can be observed that a higher-order approximation is required to maintain the same level of accuracy as the operating point R approaches $-\infty$. Nonetheless, even under this limitation, figures 4.8(b) and (c) show that the standard uncertainty estimation obtained using the analytical moment-based method, U_A, is consistently better than the estimation U_G while remaining in close agreement with the results of the MC simulation method. The approximation error increases in Test 3 shown in figure 4.8(d), which can be overcome by selecting the fifth-order approximation.

4.4 A general framework for analytical moment-based reliability and robustness analysis

As a result of using the analytical moments calculated using the Mellin transform, the outputs from the analytical moment-based standard uncertainty estimation from figure 4.1 become the inputs of the probabilistic design optimisation framework in figure 4.9. The moment-based approach to robustness analysis is implemented by finding the variance (or standard deviation) of the output Y, which is denoted by σ_Y.

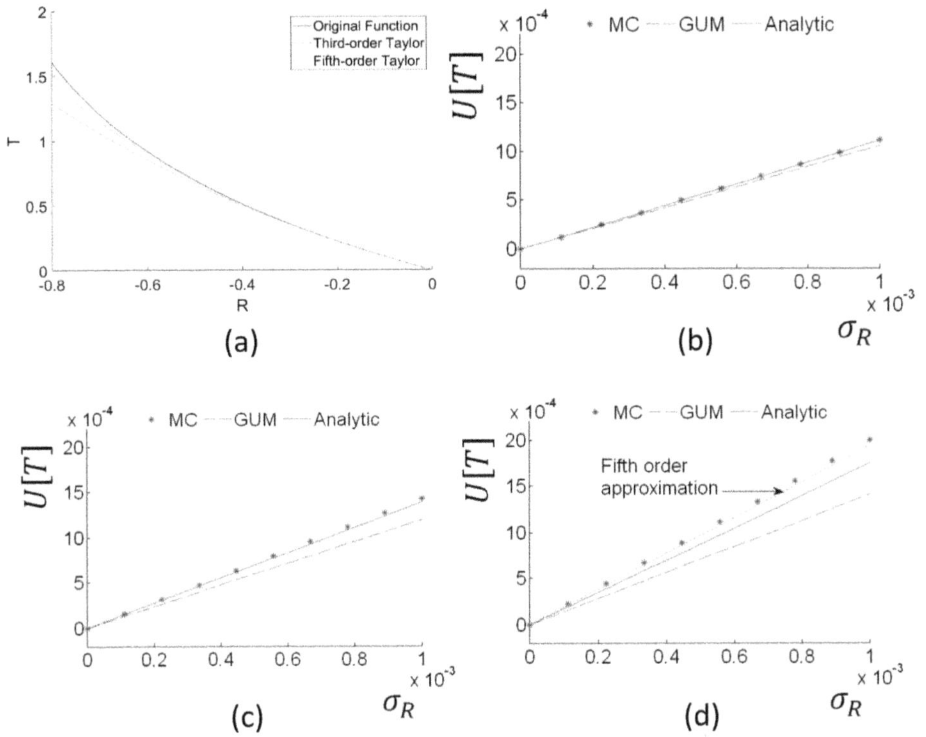

Figure 4.8. (a) Test points 1–20 in *T-R* space. (b) Test 1, (c) Test 2, and (d) Test 3 of table 4.5 through: *MC*—Monte Carlo, $U_G[\Delta f]$—*GUM*, and $U_A[\Delta f]$—*analytic* methods.

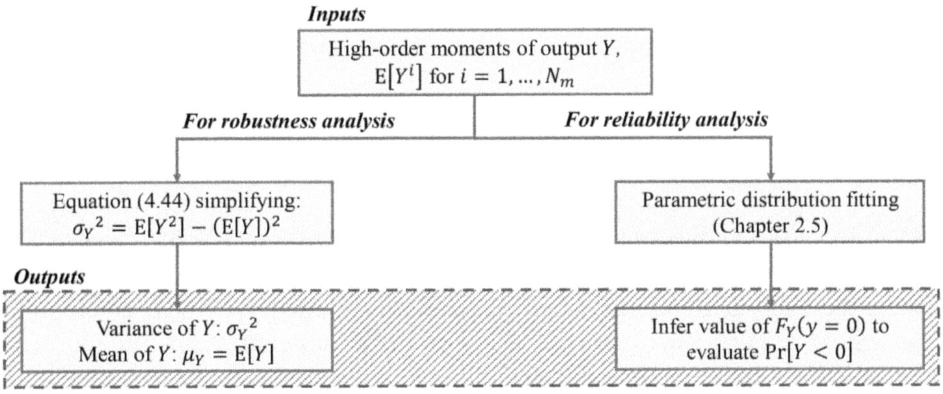

Figure 4.9. A unified framework for moment-based reliability and robustness analyses using the analytical moments derived using the Mellin transform.

Only $\mathrm{E}[Y]$ and $\mathrm{E}[Y^2]$ need to be computed, and they are represented as functions of the statistical parameters of input random variables $\mathbf{X}$. For example, assuming that Y is represented by the multivariate polynomial P, the definition of standard deviation in (2.8) from section 2.5 represents the variance as equation (4.46) by exploiting the linearity of the expectation operator and the multiplicative convolution of independent variables:

$$\sigma_Y{}^2 = \sum_{j=0}^{2N_r} a_j \left(\prod_i \mathrm{E}[X_i{}^{p_i}] \right) - \left(\sum_{j=0}^{N_r} b_j \left(\prod_i \mathrm{E}[X_i{}^{q_i}] \right) \right)^2, \tag{4.46}$$

where the indexing variables i and j, as well as the values of p_i, q_i, a_j, and b_i are kept in an abstract form, as they are not important in this discussion.

In reliability analysis, it is necessary to estimate the PDF or CDF of the output function Y using one of the distribution-fitting functions outlined in section 2.5. To accurately describe higher-order statistics such as skewness and kurtosis, it is recommended to approximate Y using at least the first four moments.

The maximum entropy distribution algorithms in section 2.5.7 are capable of computing up to the sixth-order moment of Y with a stable numerical convergence, and the use of the automated analytical moment toolboxes in section 4.2 can significantly enhance estimation accuracy while reducing the computational burden on the user. However, the choice of any fitting algorithm remains at the discretion of the user.

Once the probability distribution of Y has been estimated, the reliability analysis can be performed using the CDF. For example, the reliability $\Pr[Y < 0]$ can be directly obtained from $F_Y(y = 0)$, where y denotes the realisation of Y. The ANUNCEATOR described in section 4.2 can then be utilised to calculate the standard uncertainty and the moments up to the fourth order.

It is important to note that the general framework provided in Figure 4.9 should be customised to the intended application. Chapter 6 presents several real-world applications, each accompanied by specific probabilistic design optimisation methods.

4.5 Summary

In this chapter, a novel method for evaluating the higher-order moments of multivariate polynomials using the Mellin transform was presented. This method was applied to three case studies in the fields of magnetic force microscopy, microwave meter calibration, and eddy current measurements. The chapter also described an online toolbox, ANUNCEATOR, which allows users to easily compute the analytical moments and standard uncertainty. The chapter concluded by providing a general framework for analytical moment-based reliability and robustness analysis.

References and further reading

[1] Zhao Y-G and Ono T 2001 Moment methods for structural reliability *Struct. Saf.* **23** 47–75

[2] Lee S H and Kwak B M 2006 Response surface augmented moment method for efficient reliability analysis *Struct. Saf.* **28** 261–72

[3] Seo H S and Kwak B M 2002 Efficient statistical tolerance analysis for general distributions using three-point information *Int. J. Prod. Res.* **40** 931–44

[4] Ju B and Lee B 2008 Reliability-based design optimization using a moment method and a kriging metamodel *Eng. Optim.* **40** 421–38
[5] Zhang X and Pandey M D 2013 Structural reliability analysis based on the concepts of entropy, fractional moment and dimensional reduction method *Struct. Saf.* **43** 28–40
[6] Zhang X, Pandey M D and Zhang Y 2014 Computationally efficient reliability analysis of mechanisms based on a multiplicative dimensional reduction method *J. Mech. Des.* **136** 061006
[7] Eldred M 2009 *Recent advances in non-intrusive polynomial chaos and stochastic collocation methods for uncertainty analysis and design 50th AIAA/ASME/ASCE/AHS/ASC Structures, Structural Dynamics, and Materials Conf.* (Reston, VA: American Institute of Aeronautics and Astronautics)
[8] Lu Z-H, Cai C-H and Zhao Y-G 2017 Structural reliability analysis including correlated random variables based on third-moment transformation *J. Struct. Eng.* **143** 04017067
[9] Lu Z-H, Hu D-Z and Zhao Y-G 2016 Second-order fourth-moment method for structural reliability *J. Eng. Mech.* **143** 06016010
[10] Youn B D, Choi K K and Yi K 2005 Performance moment integration (PMI) method for quality assessment in reliability-based robust design optimization *Mech. Based Des. Struct. Mach.* **33** 185–213
[11] Lee I, Choi K K, Du L and Gorsich D 2008 Dimension reduction method for reliability-based robust design optimization *Comput. Struct.* **86** 1550–62
[12] Xu J and Lu Z-H 2017 Evaluation of moments of performance functions based on efficient cubature formulation *J. Eng. Mech.* **143** 06017007
[13] Bertrand J, Bertrand P and Ovarlez J-P 2000 The Mellin transform *The Transforms and Applications Handbook* 2 (Boca Raton, FL/Piscataway, NJ: CRC Press/IEEE Press)
[14] Mathai A M, Saxena R K and Haubold H J 2009 *The H-Function: Theory and Applications* (New York, NY: Springer)
[15] Kuang Y C, Rajan A, Ooi M P-L and Ong T C 2014 Standard uncertainty evaluation of multivariate polynomial *Measurement* **58** 483–94
[16] Springer M D and Thompson W E 1966 The distribution of products of independent random variables *SIAM J. Appl. Math.* **14** 511–26
[17] Sidi A 2003 *Practical Extrapolation Methods: Theory And Applications* (Cambridge: Cambridge University Press)
[18] Zemanian A H 1965 *Distribution Theory and Transform Analysis: An Introduction to Generalized Functions, with Applications* (New York: McGraw-Hill)
[19] Cox M G and Harris P M 2006 *Software support for metrology best practice guide no. 6: uncertainty and statistical modelling* DEM-ES 011 National Physical Laboratory https://eprintspublications.npl.co.uk/3879/
[20] Agarwal R P, Perera K and Pinelas S 2011 Cauchy's residue theorem *An introduction to Complex Analysis* (Boston, MA: Springer) pp 207–14
[21] MATHEMATICA version 13.2, 2023 (Champaign, IL: Wolfram Research Inc)
[22] MATLAB version 9.1.0.441655 (R2016b), 2016 (Natick, MA: The MathWorks Inc.)
[23] Peebles P Z, Read J and Read P 2001 *Probability, Random Variables, and Random Signal Principles* (New York: McGraw-Hill)
[24] Lomnicki Z A 1967 On the distribution of products of random variables *J. R. Stat. Soc.* B **29** 513–24
[25] Springer M D 1979 *The Algebra of Random Variables* (New York, NY: Wiley)

[26] Weisstein E W 2002 *CRC Concise Encyclopedia of Mathematics* (New York, NY: Chapman & Hall/CRC)

[27] Wold S, Esbensen K and Geladi P 1987 Principal component analysis *Chemometr. Intell. Lab. Syst.* **2** 37–52

[28] Madsen H O, Krenk S and Lind N C 2006 *Methods of Structural Safety* (Mineola, NY: Dover Publications) https://store.doverpublications.com/0486445976.html

[29] Noh Y, Choi K K and Du L 2009 Reliability-based design optimization of problems with correlated input variables using a Gaussian Copula *Struct. Multidisc. Optim.* **38** 1–16

[30] Ditlevsen O and Madsen H O 1996 *Structural Reliability Methods* (Chichester: Wiley)

[31] Schnabel R B and Eskow E 1990 A new modified Cholesky factorization *SIAM J. Sci. Stat. Comput.* **11** 1136–58

[32] Meurer A *et al* 2017 SymPy: symbolic computing in Python *PeerJ Comput. Sci.* **3** e103

[33] Sievers S *et al* 2012 Quantitative measurement of the magnetic moment of individual magnetic nanoparticles by magnetic force microscopy *Small* **8** 2675–9

[34] Guides in Metrology Committee: JCGM 2008 *Evaluation of measurement data—guide to the expression of uncertainty in measurement* (Sèvres: BIPM) https://www.bipm.org/en/committees/jc/jcgm/publications JCGM 100:2008(E)

[35] Guides in Metrology Committee: JCGM 2008 *Supplement 1 – Propagation of distributions using a Monte Carlo method (2008)* (Sèvres: BIPM) https://www.bipm.org/en/committees/jc/jcgm/publications JCGM 101:2008

[36] Ribeiro A S, Alves e Sousa J, Costa C O, Castro M P and Cox M G 2008 Uncertainty evaluation and validation of a comparison methodology to perform in-house calibration of platinum resistance thermometers using a Monte Carlo method *Int. J. Thermophys.* **29** 902–14

Melanie Po-Leen Ooi, Arvind Rajan, Ye Chow Kuang and Serge Demidenko

Chapter 5

Moment-based expanded uncertainty evaluation in design optimization

Traditionally, infinite sequences of moments were studied to find the existence and uniqueness of a solution (or distribution). However, finding the complete moment sequence is not practical in most cases. Thus, a finite moment sequence is often used in practice. The problem of finding the *cumulative distribution function* (CDF) or the PDF, given a finite sequence of moments, is known as the *truncated moment problem* [1], and these moment-constrained maximum entropy methods were discussed earlier in section 2.5.7. These algorithms were developed for better numerical convergence and stability; however, they assume that the range of the PDF is known. Unfortunately, this is not possible for most engineering systems, thus limiting their applicability for reliability analysis. Furthermore, the solution to the truncated moment problem is not unique and is highly sensitive to the moment values [2]. These challenges have led to the widespread misconception that the truncated moment approach is unreliable for the evaluation of expanded uncertainty.

To overcome these limitations, the authors of [3] proposed an improved *maximum entropy* (MaxEnt) algorithm to approximate the lower and upper limits of the distribution to ensure the convergence of the maximum entropy algorithm. This is achieved by approximating the integral limits and simplifying the multidimensional GOPoly algorithm [4] to a one-dimensional algorithm. These improvements simplify and enhance the numerical stability of the GOPoly algorithm for constraint reliability analysis in probabilistic design optimisation.

This chapter first presents the simplified and numerically stable moment-constrained maximum entropy method given in [3]. In section 5.2, a set of analytically derived benchmark test distributions is established along with a performance assessment framework for various distribution-fitting algorithms. Finally, in section 5.3, the performance assessment framework is applied with the MaxEnt algorithm along

doi:10.1088/978-0-7503-4931-4ch5

with the other mainstream distribution-fitting techniques for reliability analysis. The applications of this approach include unimodal and multimodal distributions with a wide range of skewness and kurtosis as well as different numbers of moments.

5.1 The improved moment-constrained maximum entropy method

The maximum entropy algorithm was presented earlier in section 2.5.7, which showed the Lagrange multipliers $\boldsymbol{\varphi} = \{\varphi_0, \varphi_1, \dots, \varphi_{N_m}\}$ in equation (2.27) along with their closed-form solution in equation (2.28). This subsection presents the improved moment-constrained maximum entropy algorithm with enhanced numerical stability, known as the MaxEnt algorithm, shown in figure 5.1, which is used to find these Lagrange multipliers in conjunction with with some supporting expressions provided in sections 5.1.1–5.1.3.

The MaxEnt algorithm consists of seven steps:

Step 1: Precondition the moments $\mathrm{E}[X^i]$ for $i = 0, \dots, N_m$ by setting a zero mean and unit variance (elaborated in section 5.1.1). The preconditioned moment is denoted by $\widetilde{m}_i$ and the corresponding Lagrange multiplier is $\widetilde{\varphi}_i$.

Step 2: Generate a set of random linearly independent polynomials $a_k(x)$ of the order N_m, where $k = 1, \dots, (N_m + 1)$. The Lagrange multipliers of the new basis $a_k(x)$ are denoted by γ_k. Mathematically, $\{x^i, \varphi_1\}$ for $i = 0, \dots, N_m$ is transformed to $\{a_k(x), \gamma_k\}$ for $k = 1, \dots, (N_m + 1)$.

Step 3: Using the preconditioned moments $\widetilde{m}_i$ for $i = 0, \dots, N_m$ and the new polynomial basis $a_k(x)$, calculate the starting set of Lagrange multipliers γ_k for $k = 1, \dots, (N_m + 1)$ corresponding to the moments of the standard normal distribution.

Step 4: Find the approximated integral limits (elaborated in section 5.1.2). The estimation considers the tail characteristics of the distribution based on the information provided by the moment values.

Step 5: Re-orthogonalise the set of polynomials $a_k(x)$ for $k = 1, \dots, (N_m + 1)$ using the modified Gram–Schmidt algorithm [5] by following algorithm 5.1 in section 5.1.3 for the current iteration. The orthogonal polynomial

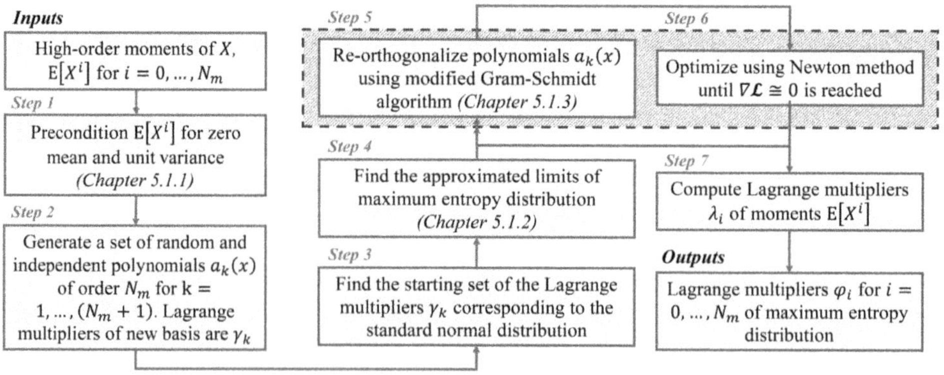

Figure 5.1. The improved moment-constrained maximum entropy algorithm with enhanced numerical stability. Steps 5 and 6 are repeated until $\nabla\mathcal{L} \cong 0$ is reached while keeping the condition number of the inverse Hessian within a threshold value.

corresponding to $a_k(x)$ is denoted by $p_k(x)$. Recompute the set of Lagrange multipliers γ_k for $p_k(x)$.

Step 6: Apply Newton's method [6] until the gradient $\nabla\mathcal{L} \cong 0$ is reached or the inverse Hessian matrix $\mathbf{H}^{-1}$ becomes too 'ill-conditioned' [4]. The inverse Hessian is deemed ill-conditioned if the condition number of the matrix $\chi_\mathbf{H}$ exceeds a threshold value (in this book, it is predefined as 20). Compute $\nabla\mathcal{L}$ and $\mathbf{H}$ using expressions (2.30) and (2.31) respectively (from section 2.5.7).

Step 7: If the gradient $\nabla\mathcal{L} \cong 0$ is reached, compute the Lagrange multipliers $\widetilde{\varphi}_i$ of the preconditioned moments $\widetilde{m}_i$ from γ_k and $p_k(x)$. Section 5.1.1 shows the method used to compute a standard set of Lagrange multipliers φ_i for the original moments $\mathrm{E}[X^i]$ from $\widetilde{\varphi}_i$. If the threshold is exceeded, i.e. $\chi_\mathbf{H} \geqslant 20$, return to step 5. The polynomials are re-orthogonalised to prevent the loss of their orthogonality, which could negatively affect the convergence of Newton's method.

The convergence of the MaxEnt algorithm shown in figure 5.1 greatly depends on the accuracy of the calculation of the integrals (2.29)–(2.31) as well as (5.5). The Gauss–Hermite quadrature rule given in [4] can be used to compute the integrals effectively. This book provides a toolbox for the MaxEnt algorithm, which is described in section 5.4.

5.1.1 Setting the zero mean and unit variance

The moments $\mathrm{E}[X^i]$ for $i = 0, \ldots, N_m$ are conditioned to a zero mean and unit variance distribution using the following transformation:

$$\widetilde{m}_i = \frac{1}{\left(\sqrt{\mathrm{E}[X^2]}\right)^i} \sum_{j=0}^{i} \binom{i}{j} (-1)^j \mathrm{E}[X^{i-j}] \mathrm{E}[X^0]^j, \tag{5.1}$$

where $\widetilde{m}_i$ denotes the transformed moment value. The inverse transformation of the Lagrange multipliers $\widetilde{\varphi}_i$ of the preconditioned moments $\widetilde{m}_i$ to the Lagrange multipliers φ_i of the original moments $\mathrm{E}[X^i]$ is given by:

$$\varphi_i = \sum_{j=0}^{n} \binom{j}{i} (-1)^{j-i} \widetilde{\varphi}_j \mathrm{E}[X^1]. \tag{5.2}$$

5.1.2 Finding integral limits

The integral limits can be approximated using the theoretical findings reported in [7]. First, use linear transformation to shift the original moments $\mathrm{E}[X^i]$ for $i = 0, \ldots, N_m$ to the point of interest C:

$$\acute{m}_i = \sum_{j=0}^{i} \binom{i}{j} (-C)^{i-j} \mathrm{E}[X^j], \tag{5.3}$$

where $\acute{m}_i$ is the linearly shifted moment.

Then, find the maximal mass ϕ at zero using equation (5.4). In this book, $\phi = 10^{-(\acute{n}+1)}$ is used to solve expression (5.4) in the real coordinate space for C. The solutions are used as the upper and lower limits of the integrals (2.29)–(2.31) and (5.5).

$$\phi = \frac{\begin{vmatrix} \acute{m}_0 & \acute{m}_1 & \cdots & \acute{m}_{\acute{n}} \\ \acute{m}_1 & \acute{m}_2 & \cdots & \acute{m}_{\acute{n}+1} \\ \vdots & \vdots & \ddots & \vdots \\ \acute{m}_{\acute{n}} & \acute{m}_{\acute{n}+1} & \cdots & \acute{m}_{2\acute{n}} \end{vmatrix}}{\begin{vmatrix} \acute{m}_2 & \acute{m}_3 & \cdots & \acute{m}_{\acute{n}+1} \\ \acute{m}_3 & \acute{m}_4 & \cdots & \acute{m}_{\acute{n}+2} \\ \vdots & \vdots & \ddots & \vdots \\ \acute{m}_{\acute{n}+1} & \acute{m}_{\acute{n}+2} & \cdots & \acute{m}_{2\acute{n}} \end{vmatrix}}, \tag{5.4}$$

where $\acute{n} = \left\lfloor \frac{N_m}{2} \right\rfloor$.

5.1.3 The modified Gram–Schmidt algorithm for polynomial orthogonalisation

The maximum entropy method proposed in [4] uses the modified Gram–Schmidt orthogonalisation method given in [8], which was designed for the polynomial basis $p_k(x)$ for $k = 1, \ldots, (N_m + 1)$. This is presented in algorithm 5.1, which uses the function $Q(\cdot)$ shown in equation (5.5).

Algorithm 5.1. An algorithm to perform modified Gram–Schmidt orthogonalisation.

Input: Linearly independent nth-order polynomials $a_k(x)$ and their Lagrange multipliers γ_k for all $k = 1, \ldots, (N_m + 1)$

Output: Orthogonalised nth-order polynomials $p_k(x)$ and their corresponding γ_k for all $k = 1, \ldots, (N_m + 1)$

Initialise the square matrix **a** of size $(N_m + 1)$, where each column vector corresponds to the coefficients of the polynomial $a_k(x)$; vector $\mathbf{g} = \{\gamma_1, \ldots, \gamma_{N_m+1}\}$ has a size of $(N_m + 1)$ corresponding to all values of γ_k; $\mathbf{g}^{initial} = \mathbf{g}$; and $\mathbf{a}^{initial} = \mathbf{a}$.
for $k = 1$ to $N_m + 1$ **do**
  for $l = 1$ to 2 **do**
    for $m = 1$ to $k - 1$ **do**
      $a_k(x) = a_k(x) - Q\left[a_k(x)p_m(x)\right]p_m(x)$.
      Update matrix **a**.
      $\mathbf{G}=\mathbf{a}^{-1}(\mathbf{a}^{\text{initial}}\mathbf{g}^{\text{initial}})$.
    end for
  end for
  $p_k(x) = Q([a_k(x)]^2)^{-\frac{1}{2}}a_k(x)$
end for
Initialise the square matrix **p** of size $(N_m + 1)$, where each column vector corresponds to the coefficients of polynomial $p_k(x)$
$\boldsymbol{\gamma} = \mathbf{p}^{-1}(\mathbf{a}^{\text{initial}}\mathbf{g}^{\text{initial}})$
return $p_k(x)$ and γ_k for all k

$$Q(g) = \int g \, \exp\left(\sum_{i=1}^{N_m+1} \gamma_i p_i(x) \right) dx, \tag{5.5}$$

where g is a multivariate polynomial.

5.2 Test distributions for benchmarking and performance analysis

Section 2.5 provided a summary of different distribution-fitting algorithms available in the literature that can be used for reliability analysis, namely *normal or Gaussian* [9], *Cornish–Fisher* (CF) [10], *extended generalised lambda distribution* (EGLD) [11], *Tukey's gh* (TGH) [12], the *Pearson system* [13, 14], and *Johnson distribution* [15, 16]. Each of the presented algorithms was tested on a different set of test distributions, thus making it challenging to evaluate their relative performances. Furthermore, many of the test distributions used in these works provided limited coverage of distributions with high skewness, high kurtosis, or both.

It is not possible to provide a fair performance comparison of different fitting techniques without a set of benchmark test distributions. This makes it, therefore, impossible to provide a recommendation for expanded uncertainty or reliability assessment of real-world systems. For example, figure 5.2 shows the output distributions (marked in red '×') used in three different research papers that investigated different fitting techniques [14, 17, 18] on a skewness–kurtosis plot. It can be seen that these distributions covered a relatively small range of skewness and kurtosis.

The study in [19] proposed a more comprehensive set of benchmark test distributions (marked as blue '○' symbols in figure 5.2), which provided wider coverage across the skewness–kurtosis plot. However, it should be noted that finite

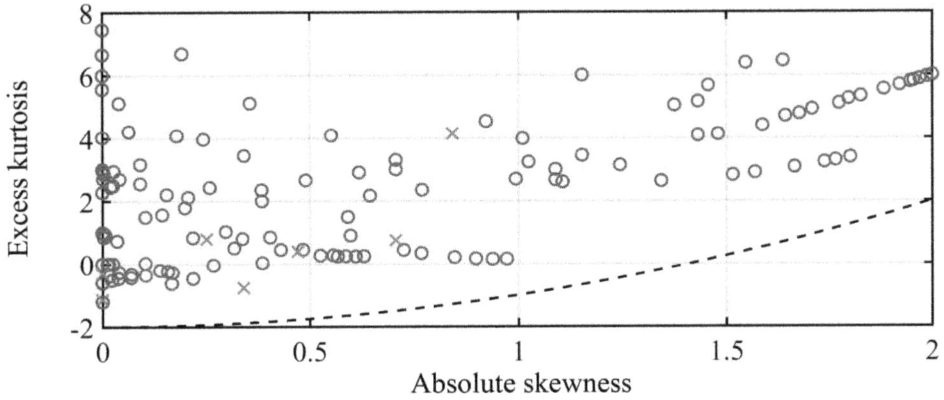

Figure 5.2. Comparisons of the skewness and kurtosis coverage in the test distributions. Points denoted by × are reported in the literature [14, 17, 18], while the points denoted by ○ represent the 130 unimodal and multimodal test distributions shown in this book. The dashed line defines the mathematical limit between skewness and kurtosis such that no distribution can occupy the region below the line [11].

moments do not determine the distribution uniquely. Hence, no matter how large the set of test distributions, it is insignificant compared to the possible space of all distributions. Therefore, the methods used to develop the test distributions are provided in sections 5.2.1–5.2.3 to support expansion efforts by readers of this book. The dashed line in figure 5.2 defines the mathematical limit which prevents distributions from occupying the region below the line [11]. These test distributions provide a common reference for anyone who intends to assess the performance of new distribution-fitting techniques or other reliability estimation frameworks.

5.2.1 Unimodal distributions

The benchmark test distributions selected for this study are those for which the confidence intervals can be analytically derived from their CDFs. Distributions generated using the Pearson system, the EGLD, Tukey's *gh*, or the CF expansion were not considered, as they would give an unfair advantage to the originating method. However, the normal and uniform distributions are included because of their practical importance.

Parametric distribution-fitting methods can adjust to changes in mean and standard deviation without affecting accuracy, so the mean and standard deviation of a distribution are not used to describe it. Instead, skewness and kurtosis are used as characterising features. If several distributions have similar skewness and kurtosis values, only one representative value is chosen. The original test distributions are listed in table 5.1; those above the dashed line are symmetrical, while the rest are skewed distributions. Additional test distributions are created by convolving a symmetrical distribution with a skewed distribution; the CDF of the new distribution is analytically derived or numerically approximated. Tables 5.2 and 5.3 present the expressions for PDF and cumulative distribution function for the original distribution set.

For example, assume that the variable X_1 follows a Weibull distribution with the shape parameter $\omega = 1$, while the variable X_2 follows an exponential distribution with the shape parameter $\omega = 2$. From tables 5.1 and 5.3, the CDF of X_1 is $f_{X_1}(x) = 1 - \exp(-x_1)$. Furthermore, the PDF of X_2 can be found from tables 5.1 and 5.2; it is $f_{X_2}(x) = 0.5\exp(-|\, x\,|)$. A new random variable is created, namely $Z = X_1 + X_2$, and the CDF of Z can then be derived using the convolution integral:

$$F_Z(z) = \int F_{X_1}\left(\frac{z - bx_2}{a}\right) f_{X_2}(x_2)\mathrm{d}x_2 \tag{5.6}$$

with $a = 1$ and $b = 1$, which gives:

$$F_Z(v) = \frac{1}{2}\left(\mathrm{erf}(z) - e^{\frac{1}{4}-z}\mathrm{erfc}\left(\frac{1}{2} - z\right) + 1\right). \tag{5.7}$$

Subsequently, the ith-order moment of Z can be analytically derived from the moment definition in equation (3.19) from chapter 3. In this example, the Weibull distribution has an absolute skewness of 2.00 and an excess kurtosis of 6.00, while the exponential distribution has an excess kurtosis of 6.00 but no skewness. The resultant distribution, however, has a skewness of 1.09 and a kurtosis of 2.67, which

Table 5.1. A list of original test distributions used. The dashed horizontal line separates the symmetrical and asymmetrical distributions. The additional test distributions are derived by convolving random variables from symmetrical distributions with another random variable from skewed distributions. The PDFs and CDFs for each distribution can be found by matching the Code to those in tables 5.2 and 5.2, respectively. Ω and Ψ_e denote the skewness and excess kurtosis, respectively.

	Distribution name and definition	Code	Ω	Ψ_e
1	Uniform distribution between [0, 1]	U01	0	−1.20
2	Symmetrical triangular distribution between −1 and 1	T	0	−0.60
3	Exponential type, shape $\omega = 3$	E3	0	−0.58
4	Normal distribution	E2	0	0.00
5	Student's t, degree of freedom $\omega = 10$	ST	0	1.00
6	Exponential type, shape $\omega = 1.1$	E11	0	2.28
7	Laplace distribution	E3	0	3.00
8	Exponential type, shape $\omega = 0.9$	E09	0	4.03
9	Exponential type, shape $\omega = 0.8$	E08	0	5.57
10	Product of two exponential-type distributions, shape $\omega = 2$	$E2^2$	0	6.00
11	Product of two exponential-type distributions, shape $\omega = 1.9$	$E19^2$	0	6.66
12	Product of two exponential-type distributions, shape $\omega = 1.8$	$E18^2$	0	7.45
13	Product of three independent triangular distributions	T^3	0	10.82
14	Weibull distribution, scale parameter $\delta = 1$, shape $\omega = 3$	W3	0.17	−0.27
15	Weibull distribution, scale parameter $\delta = 1$, shape $\omega = 2$	W2	0.63	0.25
16	Product of two independent uniform distributions	$U01^2$	0.97	0.15
17	Product of three independent uniform distributions	$U01^3$	1.80	3.39
18	Weibull distribution, scale parameter $\delta = 1$, shape $\omega = 1$	W1	2.00	6.00
19	Weibull distribution, scale parameter $\delta = 1$, shape $\omega = 0.8$	W08	2.81	12.74

falls in between its two input distributions on the skewness–kurtosis plot. This scenario is common in metrology when multiple random variables propagate through nonlinear systems with a single output.

Using this convolution method, a wider range of test distributions on the skewness–kurtosis plot was obtained. In addition to these, other test distributions were derived for $Z = aX_1 + bX_2$ using various combinations of a and b for integers within the interval [1, 4] and the convolution integral in equation (5.6). The resultant distributions were then shortlisted such that they were reasonably spread out in the region $0 \leqslant |\Omega| \leqslant 2$ and $-2 \leqslant \Psi_e \leqslant 8$ (table 5.3). This resulted in 124 test distributions as listed in table 5.4.

5.2.2 Multimodal distributions

Multimodal distributions correspond well to the operations of many real-world systems. For example, bimodal distributions are used to characterise systems such as the power loss of a V6 gasoline engine [13] and the diffusion concentrations resulting from hazardous releases [20]. Unfortunately, while unimodal distributions are widely available in the literature, multimodal test distributions can hardly be found.

Table 5.2. PDF expressions for the original distributions for $\sigma = 1$ for all distributions and parameter ω shown in table 5.1.

Code	PDF expressions
U01	$[\mathrm{H}(x) - \mathrm{H}(x-1)]$
$\mathrm{U01}^2$	$\log\left(\frac{1}{x}\right)$
$\mathrm{U01}^3$	$\frac{\left[\log\left(\frac{1}{x}\right)\right]^2}{2}$
W	$\omega(x)^{\omega-1}e^{-(x)^{\omega}}$
E	$\frac{\omega}{2\Gamma\left(\frac{1}{\omega}\right)}\exp(-\mid x\mid^{\omega})$
ST	$\frac{\Gamma\left(\frac{\omega+1}{2}\right)}{\sqrt{\omega\pi}\,\Gamma\left(\frac{\omega}{2}\right)}\left(1+\frac{1}{\omega}(x)^2\right)^{\frac{\omega+1}{2}}$
T	$(1 - \mid x\mid), \mid x\mid \leqslant 1$
T^3	$2\left(-12(-1+\mid x\mid) - 6(1+\mid x\mid)\log\left(\frac{1}{\mid x\mid}\right) - (-1+\mid x\mid)\left[2\sqrt{\mid x\mid}\right]^2\right)$
E^2	$\frac{\omega}{\Gamma^2\left(\frac{1}{\omega}\right)}K_0\left(2\left[\sqrt{\mid x\mid}\right]^2\right)$

Table 5.3. Cumulative distribution function expressions of the original distributions for $\sigma = 1$ for all distributions and parameter ω shown in table 5.1.

Code	Cumulative distribution function expressions
U01	$x[\mathrm{H}(x) - \mathrm{H}(x-1)]$
$\mathrm{U01}^2$	$x(1 - \log(x))$
$\mathrm{U01}^3$	$\frac{x(2+(-2+\log(x))\log(x))}{2}$
W	$1 - \mathrm{e}^{-(x)^{\omega}}$
E	$\frac{2\Gamma\left(\frac{1}{\omega}\right) - \Gamma\left(\frac{1}{\omega}, x^{\omega}\right)}{2\Gamma\left(\frac{1}{\omega}\right)} \quad x \geqslant 0$
	$\frac{\gamma\left(\frac{1}{\omega}, (-x)^{\omega}\right)}{2\Gamma\left(\frac{1}{\omega}\right)} \quad x < 0$
ST	$\frac{x\Gamma\left(\frac{\omega+1}{2}\right)\,{}_2F_1\left(\frac{1}{2}, \frac{\omega+1}{2}; \frac{3}{2}; -\frac{x^2}{\omega}\right)}{\sqrt{\pi}\sqrt{\omega}\,\Gamma\left(\frac{\omega}{2}\right)} + \frac{1}{2}$
T	$\frac{-x^2+2x+1}{2} \quad 0 < x \leqslant 1$
	$\frac{(x+1)^2}{2} \quad -1 < x \leqslant 0$
T^3	$\frac{1}{2}(-31x^2 + 32x + 1) - (x-2)x\log^2 x + x(7x+8)\log x \quad 0 < x \leqslant 1$
	$\frac{(x+1)^2}{2} \quad -1 < x \leqslant 0$
E^2	$\int_{-\infty}^{x} \frac{\omega K_0\left(2\mid x\mid^{\frac{\omega}{2}}\right)}{\Gamma\left(\frac{1}{\omega}\right)^2}\, dx$

Table 5.4. A list of benchmark test distributions used in the study and their statistical properties.

No.	Dist.	No.	Dist.	No.	Dist.	No.	Dist.
1	E3	32	T+W1	63	$1*E1+2*U01^2$	94	$2*E2^2 + 3*W1$
2	U01	33	E2+W3	64	$1*E1+4*U01^2$	95	$3*T^3 + 1*U01^2$
3	T	34	E2+W2	65	1*E1 + 2*W1	96	$4*E2^2 + 3*W1$
4	E2	35	$E2+U01^2$	66	1*E1 + 3*W1	97	$1*E2^2 + 4*W1$
5	S10	36	$E2+U01^3$	67	2*E1 + 3*W1	98	$4*T^3 + 1*U01^2$
6	E11	37	E2+W1	68	3*E1 + 2*W1	99	$3*E2^2 + 4*W1$
7	E1	38	S10+W3	69	4*E1 + 3*W1	100	$1*T^3 + 2*U01^2$
8	E09	39	S10+W2	70	1*E1 + 4*W1	101	$2*E2^2 + 1*W08$
9	E08	40	$S10+U01^2$	71	3*E1 + 1*W08	102	$3*E2^2 + 1*W08$
10	$E2^2$	41	$S10+U01^3$	72	3*E1 + 4*W1	103	$4*E2^2 + 1*W08$
11	$E19^2$	42	S10+W1	73	4*E1 + 1*W08	104	$3*T^3 + 2*U01^2$
12	$E18^2$	43	S10+W08	74	2*E1 + 1*W08	105	$2*T^3 + 1*W2$
13	W3	44	E1+W3	75	$3*E2^2 + 2*W2$	106	$2*T^3 + 1*U01^2$
14	W2	45	E1+W2	76	3*E1 + 2*W08	107	$2*T^3 + 1*U01^3$
15	$U01^2$	46	$E1+U01^2$	77	$3*E2^2 + 4*W2$	108	$3*T^3 + 1*U01^3$
16	$U01^3$	47	$E1+U01^3$	78	2*E1 + 3*W08	109	$4*T^3 + 1*U01^3$
17	W1	48	E1+W1	79	$3*E2^2 + 1*W1$	110	$1*T^3 + 2*U01^3$
18	E3+W3	49	E1+W08	80	4*E1 + 3*W08	111	$4*T^3 + 3*U01^2$
19	E3+W2	50	$E2^2+W3$	81	3*E1 + 4*W08	112	$3*T^3 + 2*U01^3$
20	$E3+U01^2$	51	$E2^2+W2$	82	$4*E2^2 + 1*W1$	113	$3*T^3 + 4*U01^2$
21	$E3+U01^3$	52	$E2^2+U01^2$	83	$1*E2^2 +2*W2$	114	$1*T^3 + 3*U01^3$
22	E3+W1	53	$E2^2+U01^3$	84	$1*E2^2 +3*W2$	115	$2*T^3 + 3*U01^3$
23	U01+W3	54	$E2^2+W1$	85	$2*E2^2 +3*W2$	116	$3*T^3 + 1*W1$
24	U01+W2	55	T^3+W3	86	$1*E2^2 +4*W2$	117	$4*T^3 + 3*U01^3$
25	$U01+U01^2$	56	T^3+W2	87	$3*T^3 + 1*W2$	118	$1*T^3 + 4*U01^3$
26	$U01+U01^3$	57	T^3+U01^2	88	$2*E2^2 + 1*W1$	119	$4*T^3 + 1*W1$
27	U01+W1	58	T^3+U01^3	89	$1*E2^2 + 2*W1$	120	$3*T^3 + 4*U01^3$
28	T+W3	59	T^3+W1	90	$4*T^3 + 1*W2$	121	$3*T^3 + 2*W1$
29	T+W2	60	1*E1+2*W2	91	$3*E2^2 + 2*W1$	122	$2*T^3 + 1*W1$
30	$T+U01^2$	61	1*E1+3*W2	92	$3*T^3 + 2*W2$	123	$1*E1^2 + 3*W1$
31	$T+U01^3$	62	1*E1+4*W2	93	$1*E2^2 + 3*W1$	124	$1*E1^2 + 4*W1$

The test distribution generation methodology that was presented in section 5.2.1 is not valid for the generation of multimodal distributions, because the modality of the resultant distribution cannot be guaranteed.

Instead, six additional multimodal distributions from papers [13, 20–23] are included in the set of 124 unimodal test distributions in section 5.2.1. The PDFs of the distributions are listed below in equations (5.8)–(5.13); their high-order moments were analytically determined using the moment definition given by chapter 3, equation (3.19).

$$f(x) = \frac{x^7}{3\omega(8,1)} + \frac{x^{15}}{3\omega(16,1)} + \frac{x^{63}(1-x)^{-\frac{1}{2}}}{3\omega\left(64, \frac{1}{2}\right)} \tag{5.8}$$

$$f(x) = \frac{0.5\exp\left(-\frac{(x+4)^2}{32}\right)}{\sqrt{32\pi}} + \frac{0.5\exp\left(-\frac{(x-4)^2}{18}\right)}{\sqrt{18\pi}} \tag{5.9}$$

$$f(x) = \frac{0.4\exp\left(-\frac{(x+1)^2}{0.32\sigma^2}\right) + 0.6\exp\left(-\frac{(x-1)^2}{0.32\sigma^2}\right)}{\sqrt{0.32\pi}} \tag{5.10}$$

$$f(x) = \frac{0.4\exp\left(-\frac{x^2}{2}\right) + 0.6\exp\left(-\frac{(x-3)^2}{2}\right)}{\sqrt{2\pi}} \tag{5.11}$$

$$f(x) = \frac{0.4\exp\left(-\frac{x^2}{2}\right) + 0.6\exp\left(-\frac{(x-25)^2}{2}\right)}{\sqrt{2\pi}} \tag{5.12}$$

$$f(x) = \frac{(x+1)^{23}(1-x)^{11} + (x+1)^{11}(1-x)^{23}}{2^{36}\omega\left(64, \frac{1}{2}\right)} \tag{5.13}$$

5.2.3 A performance assessment framework that uses the benchmark test distributions

The performance assessment framework presented in figure 5.3 utilises the CDFs and moments of the test distributions shown in table 5.4 and equations (5.8)–(5.13). These are obtained by deriving the analytical solution for a set of distributions in the ranges $0 \leqslant |\Omega| \leqslant 2$ and $1 \leqslant \Psi_e \leqslant 11$. These test distributions, which include unimodal and multimodal distributions, are represented by the circles in figure 5.2. For convenience, these test distributions together with their analytical solutions can be downloaded from http://polymoment.com/.

Although only moment-based parametric distribution fitting is reported in this monograph, it should be noted that the benchmark distributions with known CDFs can also be used to assess any reliability estimation technique that conforms to the classical probabilistic framework.

The framework shown in figure 5.3 is used in four simple steps:

1. A test distribution is selected from the set of benchmark test distributions.
2. The required information is derived from the PDF. This book advocates moment-based approaches. Therefore, it directly extracts the expressions for the high-order moments of the selected test distribution. This enables a trustworthy assessment of the fitting techniques for reliability analysis.

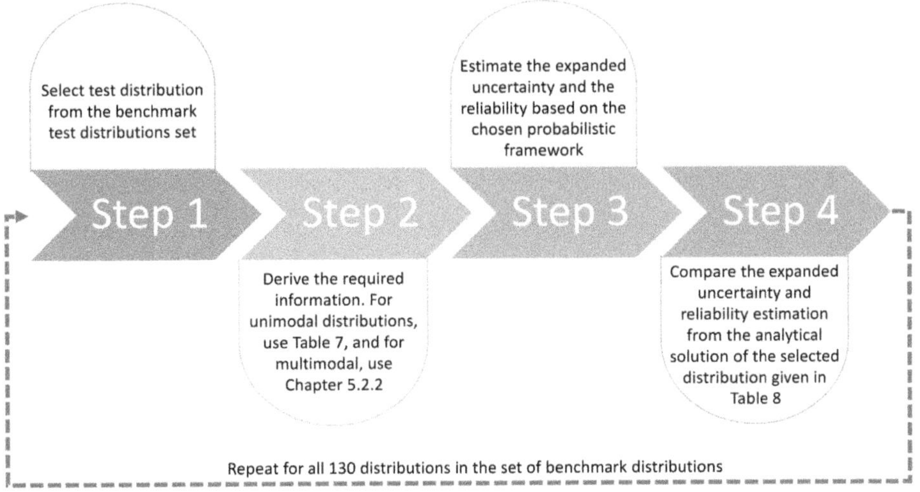

Figure 5.3. A framework for analysing the performances of reliability estimation techniques. The formulation strategy and a list of the test distributions are presented in sections 5.2.1 and 5.2.2.

3. The reliability is estimated from the expanded uncertainty using the technique under test. For the moment-based technique, the obtained moments are fed into the maximum entropy method in section 5.1 as well as the distribution-fitting techniques listed in section 2.5. Once the distribution is obtained, the reliability is estimated.
4. The estimated reliability is then compared against the actual reliability values derived from the analytical CDF solution for the selected distribution.

In this study, steps 1–4 are repeated for all 130 benchmark test distributions.

The reliability estimation performance can be then be measured using two different methods. Given that x is the realisation of X, the estimation performance can be determined via the discrepancy in the value of x for a given percentile level, which can be calculated using the relative error:

$$\varepsilon = \frac{\frac{\left|(x-\mu_X)-(x'-\mu_X)\right|}{\sigma}}{\frac{(x-\mu_X)}{\sigma X}} = \frac{|x - x'|}{x - \mu_X}, \tag{5.14}$$

where x' denotes the estimated value, x denotes the actual value, σ_X denotes the standard deviation of X, and μ_X denotes the mean of X. Since the reported values of x and x' are always normalised with respect to σ_X, ε can also be viewed as a percentage error in the *coverage factor* [24, 25] estimation.

It is reasonable to expect that reliability estimation concerns the large and small percentiles which correspond to the tails of a distribution. In this respect, the reliability evaluation is different from the standard distribution fitting, which requires the fitting to be good over the full range of the input space. In fact, some distribution-fitting applications specifically emphasise the importance of a good fit around the mean.

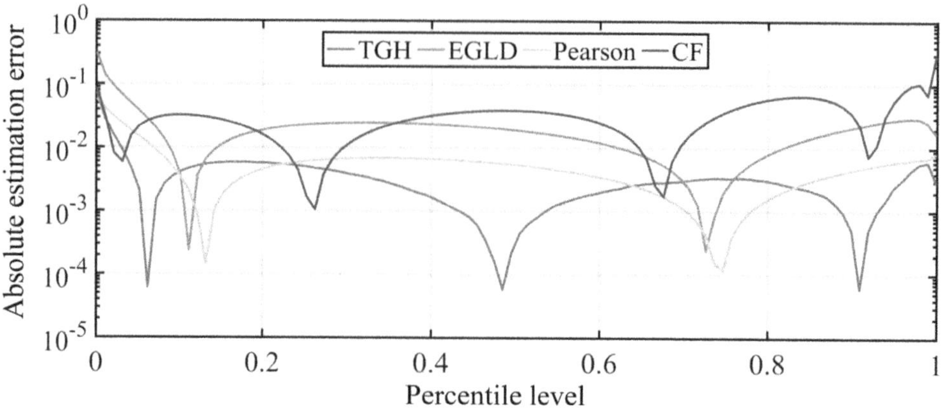

Figure 5.4. The approximation error of x for a Weibull distribution with shape parameter $\omega = 2$ fitted using the EGLD, Pearson, CF, and TGH fitting techniques.

For example, figure 5.4 shows the x value estimation error across the entire percentile level of the Weibull distribution with the shape parameter $\omega = 2$. The right and left tails of a distribution are interpreted as the percentiles higher than 90% and lower than 10%, respectively, in figure 5.4. Note that the estimation error of x is higher at the high and low percentile levels. This means that even though a distribution-fitting technique may show a good fit over the full distribution range, it may not necessarily perform well in the reliability estimation.

Since the estimation accuracy is concentrated at the tails of the distribution, two estimation metrics are defined as follows: ε_{90} reports the mean error of the x value estimation in the 90% to 99.999% range, while ε_{10} reports the mean error of the x value estimation in the 0.001% to 10% range. The performance metric can be further consolidated as $\varepsilon_{\max}$:

$$\varepsilon_{\max} = \max(\varepsilon_{10}, \varepsilon_{90}). \quad (5.15)$$

5.3 Reliability analysis of parametric distribution-fitting techniques: from unimodal to multimodal distributions

In this section, the performance assessment framework and benchmark test distributions from section 5.2 are implemented together with the following distribution-fitting methods: normal, Cornish–Fisher expansion, extended generalised lambda, Tukey's *gh*, Pearson, Johnson, and the improved maximum entropy from section 5.1. They are denoted as Normal, CF, EGLD, TGH, Pearson, Johnson, and MaxEnt respectively in the subsequent subsections.

To provide a comprehensive discussion, the performance assessment is divided as follows:

- Unimodal distributions with four moments
- Unimodal distributions with more than four moments
- Multimodal distributions

Furthermore, following the assessment results, section 5.3.4 proceeds to recommend a reliable region of application for some Tukey's *gh* and the Cornish–Fisher techniques.

5.3.1 Unimodal distributions with four moments

Consider only the 124 unimodal distributions shown in section 5.2.1. First, the exact high-order moments up to $N_m = 4$ are calculated using the analytical expressions for the 124 distributions. Then, seven sets of distributions are obtained using the Normal, CF, EGLD, TGH, Pearson, Johnson, and MaxEnt distributions. The distributions are then used to estimate the values of ε_{10}, ε_{90}, and subsequently ε_{max} using the actual values of x computed from the analytical distributions.

Figure 5.5 shows a scatter plot for the error metric ε_{max} with respect to absolute skewness and excess kurtosis for the aforementioned seven sets of distributions. The colour bar in the figure indicates the range of values for ε_{max} in terms of percentage. The MC simulation method is shown in figure 5.5(a), which consistently provided the lowest ε_{max}. Note that this method does not provide an analytical expression of uncertainty. It is used as a benchmark for all other distribution-fitting methods.

The normal approximation method in figure 5.5(g) is well-known to be reliable for normal and almost-normal distributions, or in other words, distributions that fall around the origin of the plot. It can be observed that ε_{max} becomes 'dark red', which represents 100%, particularly towards the upper right, which represents both high absolute skewness and high excess kurtosis.

Perhaps the most surprising result is the performance of the EGLD, which is a popular distribution-fitting method and is advocated by some researchers for use in reliability estimation. Figure 5.5(h) shows that its overall ε_{max} is worse than that of the normal approximation technique, and it struggles even with mildly skewed distributions. Referring to chapter 2's equations (2.12) and (2.13), the EGLD is designed to fit the tails of distributions using the λ_3 and λ_4 parameters. However, it appears that the form of the parametric equation in the EGLD is unable to appropriately model the power-law behaviours or compact support distributions, leading to its dismal performance on the benchmark test distributions.

The performances of the Johnson and the Pearson distributions are presented in figures 5.5(a) and (b). Their estimation errors ε_{max} are consistently lower than those produced by the MaxEnt distribution-fitting method. Figure 5.6 presents the results for the Johnson, Pearson, MaxEnt, TGH, and CF techniques to clearly distinguish their performances by rescaling the colour bar to 50%. It can be seen that there are many instances in which the approximation error drops to almost zero due to good approximation.

It should be noted that the Johnson and Pearson methods are systems of distributions (refer to sections 2.5.5 and 2.5.6). The MaxEnt method, on the other hand, does not have a set of distributions to choose from, unlike the Johnson and Pearson methods. Yet despite this stark difference, figure 5.5(c) shows that the MaxEnt method can perform at a level comparable to those of the Johnson and Pearson distributions for reliability estimation. The MaxEnt estimation worsens when

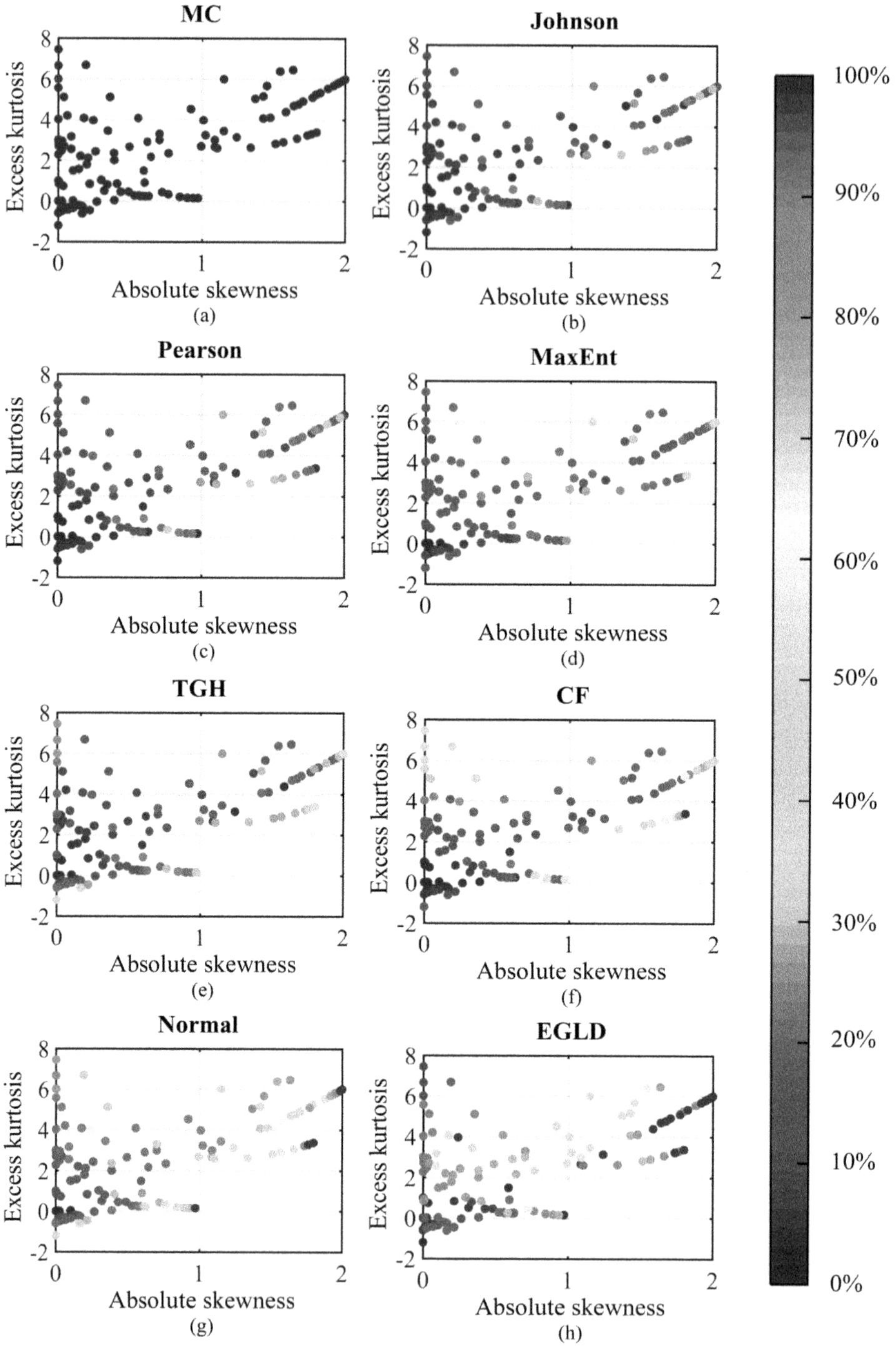

Figure 5.5. Scatter plots of $\varepsilon_{\max}$ with respect to absolute skewness and excess kurtosis for the (a) MC, (b) Johnson, (c) Pearson, (d) MaxEnt, (e) TGH, (f) CF, (g) Normal, and (h) EGLD techniques. The colour bar shows the range of values for $\varepsilon_{\max}$.

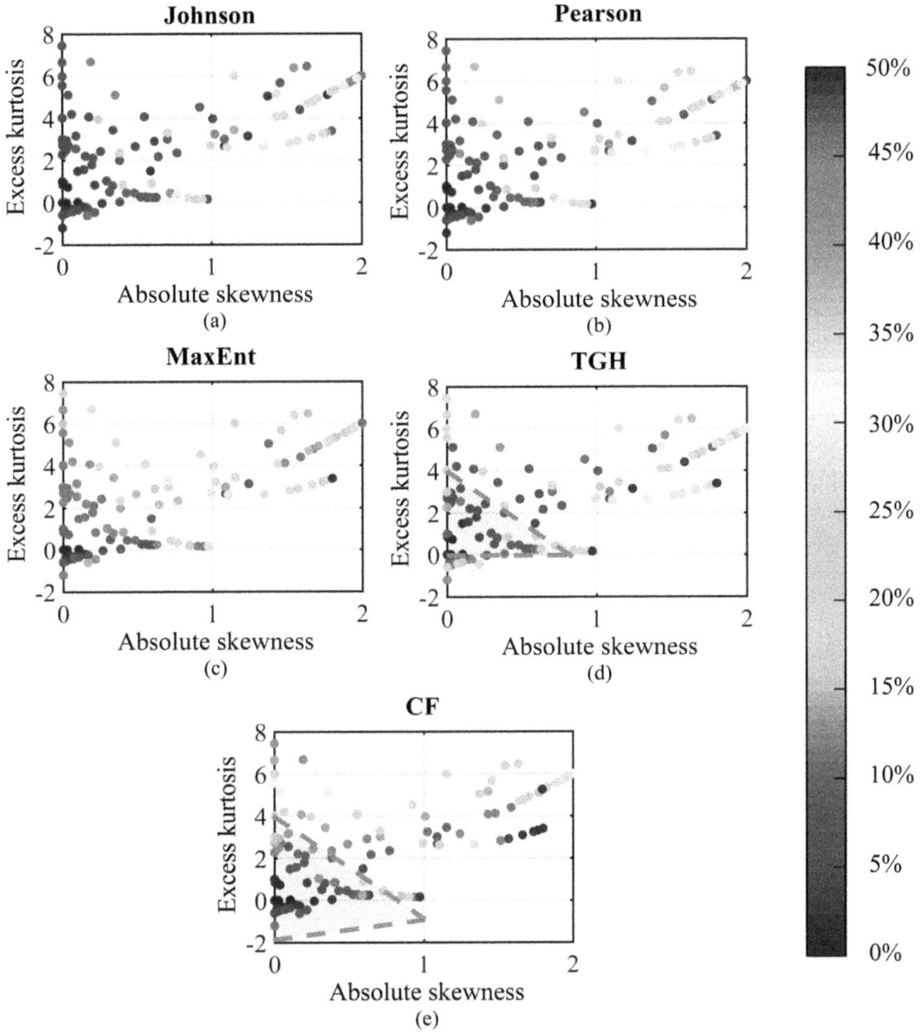

Figure 5.6. Scatter plots of $\varepsilon_{\max}$ with respect to absolute skewness and excess kurtosis for the (a) Johnson, (b) Pearson, (c) MaxEnt, (d) TGH, and (e) CF techniques. The rescaled colour bar shows the range of values for $\varepsilon_{\max}$. The red transparent region encompasses distributions for which the TGH and CF techniques can be reliably applied within a 20% error range.

the distributions have increasing skewness and increasing kurtosis. This is a somewhat expected result because only four moments are considered in this subsection, which limits the information for the tails of the distribution. It is expected that the true potential of the MaxEnt method could be demonstrated by using an increased number of moments, which will be discussed in the following subsection.

The TGH and CF were designed to modify the normal distribution based on additional information given by the third and fourth moments (refer to sections 2.5.2 and 2.5.4 respectively for their background). With such an approach, it is expected that these two techniques should outperform the normal approximation when the distribution deviates slightly from the normal. Figures 5.5(e) and (f) show that the TGH and CF algorithms offer significant and consistent improvement over the normal approximation, especially for highly skewed distributions. However, further observation of figures 5.6(d) and (e) shows that there is a region of reliable reliability estimation within 20% of error for these two methods which covers most of the symmetrical and slightly skewed distributions while disregarding distributions with high kurtosis. Following on from this observation, section 5.3.4 presents the conditions under which the TGH and CF can be reliably applied.

The skewness–kurtosis plots shown in figures 5.5 and 5.6 allow for a qualitative discussion of the estimation performances of each type of distribution-fitting method. However, these results do not reflect the consistency of the reliability estimation techniques in a quantifiable manner. To do so, a confidence plot is required, which is determined by the number of distributions falling within the intervals of reliability estimation error. In this confidence plot, the faster the line reaches the maximum, the more dependable the corresponding technique is.

Figures 5.7(a)–(c) show confidence plots for ε_{10}, ε_{90}, and ε_{max}, respectively. As expected from the previous results, the Johnson and Pearson distribution-fitting

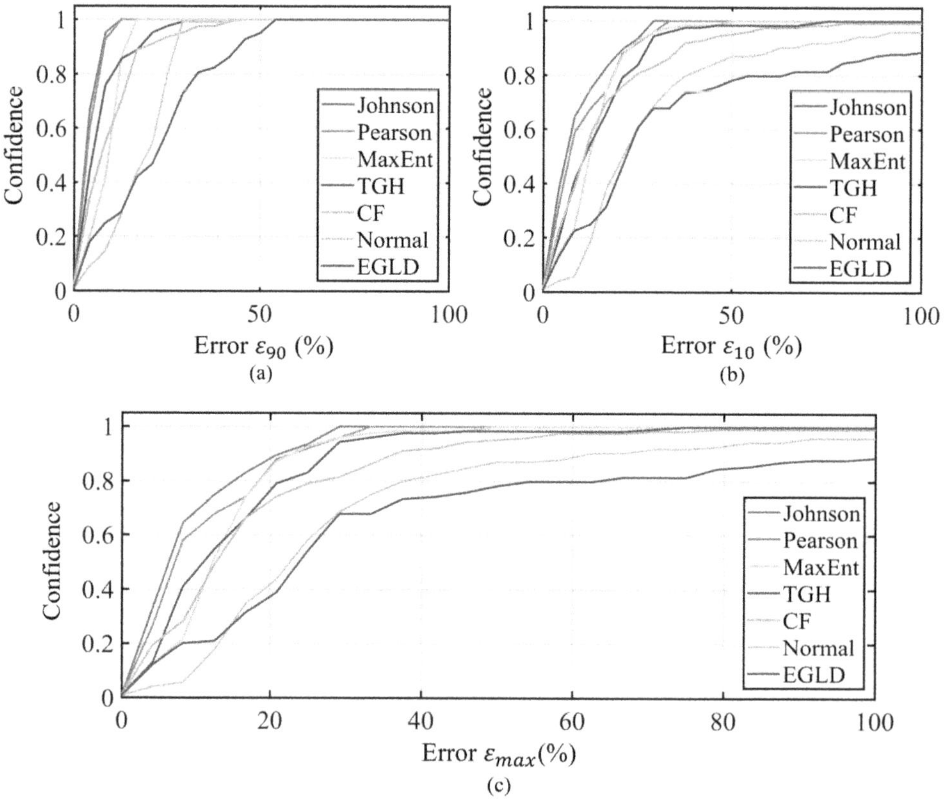

Figure 5.7. Reliability estimation confidence plots for (a) ε_{90}, (b) ε_{10}, and (c) ε_{max} obtained using the Johnson, Pearson, MaxEnt, TGH, CF, normal, and EGLD techniques. The fastest technique to reach the maximum reliability is the most dependable one.

methods shown in figure 5.7(a) and (b) are the most dependable for reliability estimation at both ends of the distribution 'tails', closely followed by the MaxEnt method. The TGH and CF techniques produce comparable results to those of the TGH method and slightly outperform the CF method. Figures 5.7(a) and (b) show that the EGLD performs worse than the normal approximation method. Therefore, this book recommends that the EGLD should not be used for reliability estimation.

Figure 5.7(c) is useful in ranking the distribution-fitting techniques. For example, at 90% confidence, the average errors $\varepsilon_{\max}$ obtained using the TGH, CF, and normal techniques are 30%, 35% and 60%, respectively, while the error of the EGLD method cannot be deduced, as it exceeds 100% in figure 5.7(c). Interestingly, the confidence plots in figure 5.7(b) stabilise relatively faster than the plots in figure 5.7(a), indicating that the x value estimation for the upper tail section is more reliable than for the lower tail section.

Many distributions do not have the same tail characteristics at both ends. For example, the lower tail could be a finite support, while the upper tail could be an infinite support. Thus, modelling such a distribution with an infinite support distribution at both ends would be advantageous for the upper tail, but not for the lower one. This would result in a less accurate reliability estimation for the lower tail, leading a higher magnitude of error compared to that of the upper tail.

5.3.2 Unimodal distributions with more than four moments

The previous subsection ultimately limited the performance of the MaxEnt algorithm because of the use of just four moments. Since the MaxEnt algorithm uses a relatively new analytical moments toolbox, it is able to accommodate higher-order moments without burdening the user. Accordingly, the exact high-order moments up to $N_m = 12$ were calculated from the analytical expressions of the 124 distributions. Then, five sets of 124 MaxEnt distributions based on $N_m = 4, 6, 8, 10, 12$ moments, respectively, were obtained from the MaxEnt algorithm.

Figure 5.8 presents a scatter plot on the skewness–kurtosis plane. Its colour bar is limited to 30% error for better clarity. Figure 5.9 presents the reliability plots, which are similar to the plots presented in the previous subsection. Table 5.5 presents the means and standard deviations of $\varepsilon_{\max}$ for the 124 test distributions. In addition, the table also reports the average computation time for each distribution (except for the Johnson and Pearson methods, as they do not require an optimisation algorithm for their deployment). Note that all computations were performed by a computer with a 64-bit Intel Core i5-3470 CPU and 8 GB of RAM.

As anticipated, the scatter plots in figures 5.8(c)–(g) show that reliability assessment using the MaxEnt algorithm improves with the number of moments, i.e. with increasing information about the tails of the distributions. Figure 5.9 and table 5.5 further validate this outcome. It can be observed from figure 5.9 that the confidence plot reaches the maximum value faster for increasing numbers of moments.

Table 5.5 shows that the means and standard deviations of $\varepsilon_{\max}$ for the MaxEnt method improve with the order of moments. However, the same table shows that

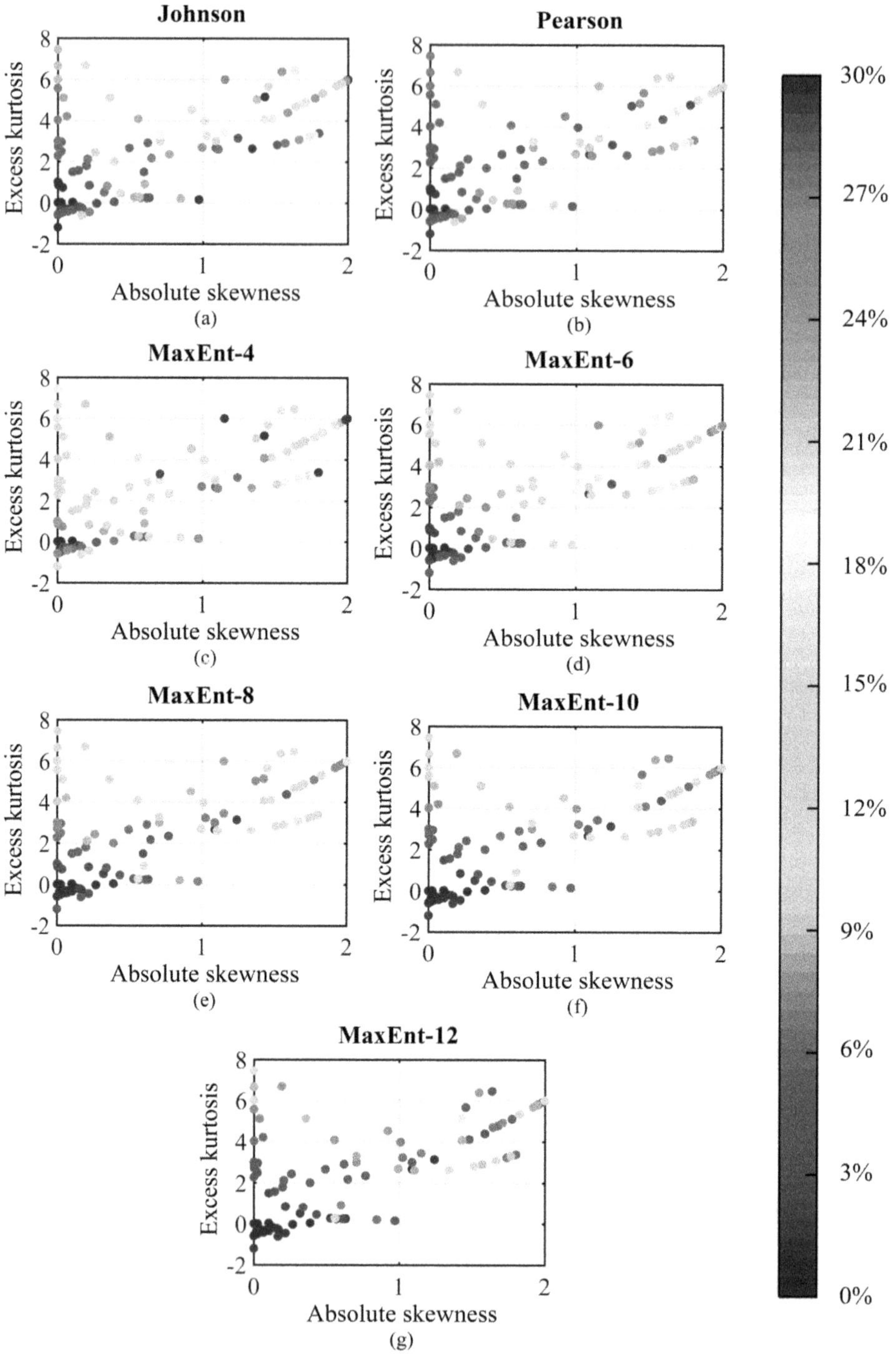

Figure 5.8. Scatter plots of $\varepsilon_{\max}$ with respect to absolute skewness and excess kurtosis for the (a) Johnson, (b) Pearson, (c) MaxEnt-4, (d) MaxEnt-6, (e) MaxEnt-8, (f) MaxEnt-10, and (g) MaxEnt-12 methods. The colour bar shows the range of values for $\varepsilon_{\max}$.

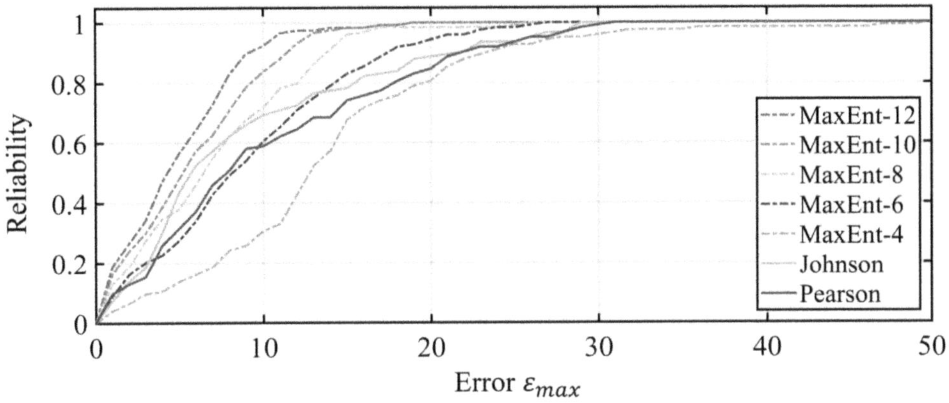

Figure 5.9. Reliability estimation confidence plots of $\varepsilon_{\max}$ produced using the Johnson, Pearson, and MaxEnt methods (for the MaxEnt method, $N_m = 4, 6, 8, 10, 12$). The fastest technique to reach the maximum reliability is the most reliable one.

Table 5.5. Mean and standard deviations of the reliability estimation errors and average computation times of the distribution-fitting algorithms based on 124 unimodal test distributions.

Method	Mean (%)	Standard deviation (%)	Average computation time (s)
Johnson	7.23	9.01	–
Pearson	8.22	11.30	–
MaxEnt—4 moments	10.99	12.27	0.4157
MaxEnt—6 moments	6.99	8.84	1.2676
MaxEnt—8 moments	5.63	7.32	2.8269
MaxEnt—10 moments	4.43	5.73	43.6925
MaxEnt—12 moments	3.72	4.98	63.6472

these performance enhancements come at the expense of additional computational time. For example, the average $\varepsilon_{\max}$ using *MaxEnt—6 moments* is 6.99%, while the average $\varepsilon_{\max}$ *MaxEnt—12 moments* is 3.72%, which comes at the cost of 50 times the computational delay, i.e. 0.4157 s versus 63.6472 s. Nevertheless, processor design and technology continue to advance and with them comes increasingly faster computational speed. Thus, this trade-off continues to reduce in severity with the advancement of technologies such as more efficient numerical integration methods for higher N_m, better computer hardware, precompiled codes, parallel processing, and different programming platforms.

5.3.3 Multimodal distributions

Most mainstream reliability estimation techniques focus on unimodal distributions because multimodal distributions are comparatively rarer. Although less frequently encountered, they are still relevant, especially in the cases described in section 5.2.2. It should be noted that the MaxEnt algorithm is capable of handling multimodal

Table 5.6. The means and standard deviations of the reliability estimations and average computation times of the distribution-fitting algorithms based on six multimodal test distributions.

Method	Mean (%)	Standard deviation (%)	Average computation time (s)
Johnson	10.12	9.74	–
Pearson	11.46	10.77	–
MaxEnt—4 moments	3.55	4.20	1.3768
MaxEnt—8 moments	1.00	1.73	3.3671
MaxEnt—12 moments	0.34	0.58	72.8133

distributions because it does not impose an assumption of unimodality, unlike the other methods.

The PDFs of the multimodal test distributions are listed in section 5.2.2; their high-order moments were calculated analytically using equation (3.19). The reliability estimation error $\varepsilon_{\max}$ was then obtained using equation (5.15). Table 5.6 presents the means and standard deviations of $\varepsilon_{\max}$ as well as the average computation time for the approximation of the six multimodal test distributions. To keep the discussion succinct, the proposed MaxEnt algorithm was employed with $N_m = 4,8,12$, and its performances were benchmarked against the results obtained using the Johnson and Pearson methods.

With four moments, i.e. $N_m = 4$, the Johnson and Pearson distribution methods, which were the best for unimodal distributions, already show less promise compared to *MaxEnt—4 moments* when they are used estimate the multimodal distributions. Table 5.6 shows that MaxEnt method for multimodal distributions becomes increasingly accurate as the number of moments increases—the average error $\varepsilon_{\max}$ falls to 0.34%. The computation time for the MaxEnt method increases with the number of moments, which is similar to the observations made for the unimodal distributions.

It is important to note that the mean and standard deviation values of $\varepsilon_{\max}$ for multimodal distributions are lower than those of the unimodal distributions (from table 5.5) because of the inclusion of challenging distributions with extreme skewness and kurtosis in the unimodal benchmark set. No such distributions are present in the multimodal benchmark set.

5.3.4 Reliable regions for Tukey's *gh* method and the Cornish–Fisher technique

Figures 5.6(d) and (e) from the earlier subsections show that Tukey's *gh* and the Cornish–Fisher technique are unstable for distributions with high skewness, high kurtosis, or both. However, a reliable region for the use of Tukey's *gh* and the Cornish–Fisher technique for uncertainty estimation can be defined based on their performances in the benchmark test distribution set. For example, Tukey's *gh* can be reliably applied for distributions with parameters of absolute skewness $| \Omega |$ and excess kurtosis Ψ_e that adhere to the following constraints:

$$\Psi_e \geqslant 0 \text{ and } \Psi_e + 5 | \Omega | \leqslant 4. \tag{5.16}$$

This limits the range of applicability of Tukey's *gh* technique for reliability estimation. For ease of use, a lookup table is presented in appendix A. It is ready to be used by practitioners to solve for the parameters g and h, provided that the parameters $| \Omega |$ and Ψ_e fulfil the conditions above (5.16).

Due to the limitation of Tukey's *gh* for distributions with shorter tails than those of the normal distribution, Tukey's method is only reliable for distributions that have positive excess kurtosis. The Cornish–Fisher method does not have such a limitation; thus, it can be used on distributions that have negative excess kurtosis. Consequently, the condition for the uncertainty estimation using the Cornish–Fisher technique is:

$$\Psi_e + 5 | \Omega | \leqslant 4. \tag{5.17}$$

5.4 A toolbox for the MaxEnt algorithm

This section presents a toolbox that can be used to compute the MaxEnt PDF using equation (2.28) given a set of raw moments. The toolbox, shown in figure 5.10 and accessible at http://polymoment.com/Maximum_Entropy_Distribution uses

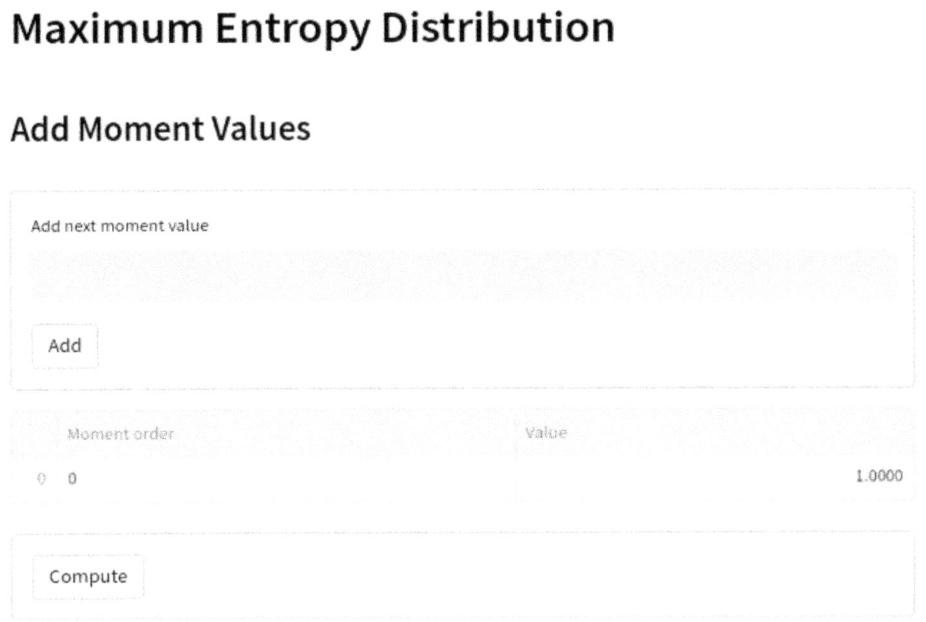

Figure 5.10. The Maximum Entropy (MaxEnt) toolbox used to compute the MaxEnt PDF from raw moment values.

SymPy [26], SciPy [27], and NumPy [28], which are Python libraries for symbolic mathematical computation, optimisation, and array operations respectively, to compute the optimum Lagrange multipliers in equation (2.28) with respect to the supplied moment values.

To get started, simply enter the raw moment values into the provided table. Note that at least two moment values must be supplied, and the zeroth moment value is always set to 1.0 (see figure 5.10). Once the moment values have been entered, click on the 'Compute' button to initiate the calculation of the distribution limits and the optimised Lagrange multipliers as defined by equation (2.28). The results will be displayed in a neatly formatted JSON file, and the corresponding PDF distribution will be plotted for reference. This can help the user validate the shape of the resulting PDF against the values of the moments.

For example, figure 5.11 shows an instance in which the moment values of a Weibull distribution have been entered into the table up to the eighth order: [1.0, 0.8862, 1.0, 1.329, 2.0, 3.3234, 6.0, 11.6317, 24.0]. When the user clicks on the 'Compute' button, the MaxEnt toolbox calculates the optimum values of

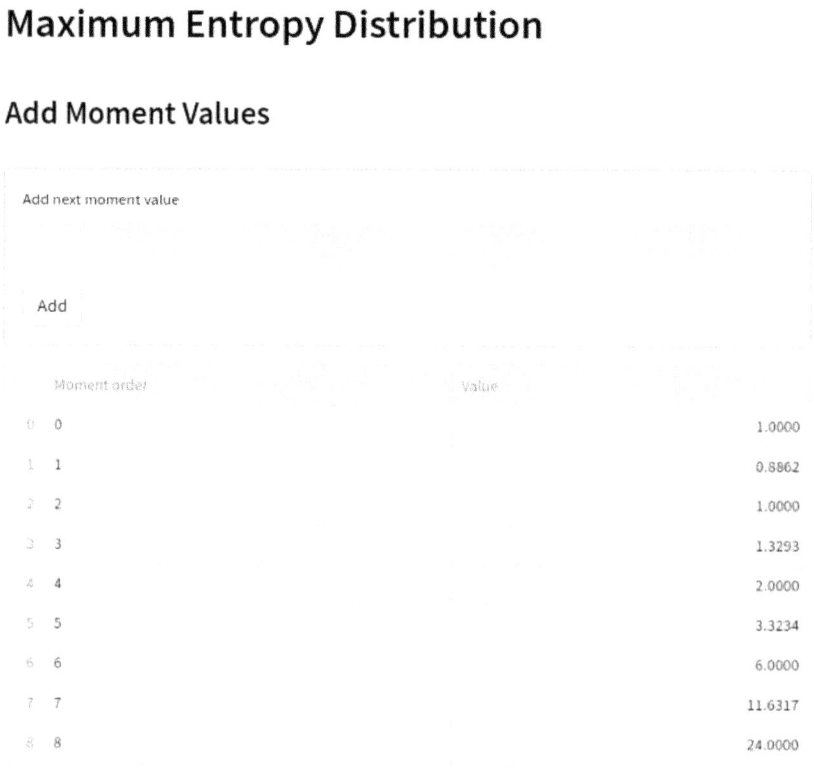

Figure 5.11. The MaxEnt toolbox; moment values for a Weibull distribution have been supplied up to the eighth order.

the Lagrange multipliers and displays the results as shown in figure 5.12(a). The distribution limits and Lagrange multipliers are also shown under the keys 'bounds' and 'phis', respectively, in figure 5.12(b), which can be copied and pasted into a different program of choice to plot the same distribution.

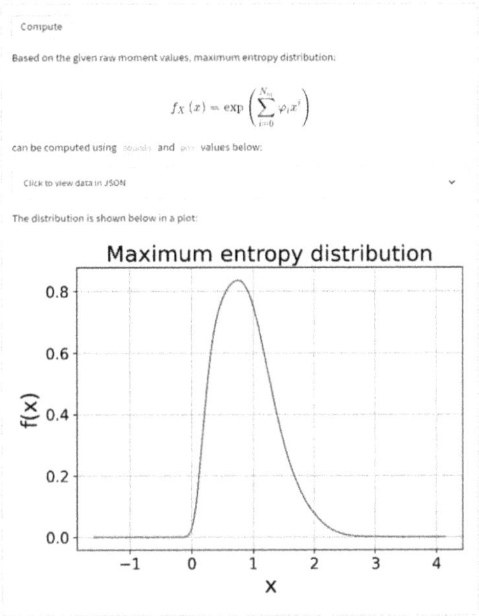

(a)

(b)

Figure 5.12. Example output from the MaxEnt toolbox: (a) is a plot of the PDF and (b) shows the bounds and Lagrange multipliers of the MaxEnt distribution given in (2.28).

5.5 Summary

This chapter presented a moment-constrained maximum entropy method that approximates the lower and upper limits of the distribution for expanded uncertainty evaluation. This method overcomes the limitations of traditional truncated moment approaches, providing a simplified and numerically stable algorithm for probabilistic design optimisation. The chapter included a performance assessment framework for various distribution-fitting algorithms and introduced a toolbox for computing the MaxEnt algorithm, which can be applied to unimodal and multi-modal distributions with a range of skewness and kurtosis.

References

[1] Lindsay B G and Basak P 2000 Moments determine the tail of a distribution (but not much else) *Am. Stat.* **54** 248–51

[2] McCullagh P 1994 Does the moment-generating function characterize a distribution? *Am. Stat.* **48** 208

[3] Rajan A, Kuang Y C, Ooi M P-L, Demidenko S N and Carstens H 2018 Moment-constrained maximum entropy method for expanded uncertainty evaluation *IEEE Access* **6** 4072–82

[4] Abramov R V *et al* 2010 The multidimensional maximum entropy moment problem: a review of numerical methods *Commun. Math. Sci.* **8** 377–92

[5] Royden H L 1953 Bounds on a distribution function when its first *n* moments are given *Ann. Math. Stat.* **24** 361–76

[6] Bertsekas D P 1999 *Nonlinear Programming* 2 (Belmont, MA: Athena Scientific)

[7] Racz S, Tari A and Telek M 2006 A moments based distribution bounding method *Math. Comput. Modell.* **43** 1367–82

[8] Giraud L, Langou J and Rozloznik M 2005 The loss of orthogonality in the Gram–Schmidt orthogonalization process *Comput. Math. Appl.* **50** 1069–75

[9] Fisher S R A and Cornish E A 1960 The percentile points of distributions having known cumulants *Technometrics* **2** 209–25

[10] Cornish E A and Fisher R A 1938 Moments and cumulants in the specification of distributions *Rev. Int. Stat. Inst.* **5** 307–20

[11] Karian Z A and Dudewicz E J 2000 *Fitting Statistical Distributions: The Generalized Lambda Distribution and Generalized Bootstrap Methods* (New York, NY: Chapman and Hall/CRC)

[12] Tukey J W 1977 Modern techniques in data analysis *Proc. of the NSF-Sponsored Regional Research Conf.* 7 *(Southern Massachusetts University, North Dartmouth, MA)*

[13] Xi Z, Hu C and Youn B D 2012 A comparative study of probability estimation methods for reliability analysis *Struct. Multidiscip. Optim.* **45** 33–52

[14] Sim C H and Lim M H 2008 Evaluating expanded uncertainty in measurement with a fitted distribution *Metrologia* **45** 178

[15] Johnson N L 1949 Systems of frequency curves generated by methods of translation *Biometrika* **36** 149–76

[16] Johnson N L, Kotz S and Balakrishnan N 1994 *Continuous Univariate Distributions* 1 2 (Hoboken, NJ: Wiley) https://www.wiley.com/en-gb/Continuous+Univariate+Distributions%2C+Volume+1%2C+2nd+Edition-p-9780471584957

[17] Lampasi D A, Di Nicola F and Podest 2006 Generalized lambda distribution for the expression of measurement uncertainty *IEEE Trans. Instrum. Meas.* **55** 1281–7
[18] Lampasi D A 2008 An alternative approach to measurement based on quantile functions *Measurement* **41** 994–1013
[19] Rajan A, Kuang Y C, Ooi M P-L and Demidenko S N 2016 Benchmark test distributions for expanded uncertainty evaluation algorithms *IEEE Trans. Instrum. Meas.* **65** 1022–34
[20] Gotovac H and Gotovac B 2009 Maximum entropy algorithm with inexact upper entropy bound based on Fup basis functions with compact support *J. Comput. Phys.* **228** 9079–91
[21] Kang H Y and Kwak B M 2009 Application of maximum entropy principle for reliability-based design optimization *Struct. Multidiscip. Optim.* **38** 331–46
[22] Li G and Zhang K 2011 A combined reliability analysis approach with dimension reduction method and maximum entropy method *Struct. Multidiscip. Optim.* **43** 121–34
[23] Provost S B 2005 Moment-based density approximants *Mathematica J.* **9** 727–56
[24] Guides in Metrology Committee: JCGM 2008 *Supplement 1 – Propagation of distributions using a Monte Carlo method (2008)* (Sèvres: BIPM) https://www.bipm.org/en/committees/jc/jcgm/publications JCGM 101:2008
[25] Guides in Metrology Committee: JCGM 2008 *Evaluation of measurement data—guide to the expression of uncertainty in measurement* (Sèvres: BIPM) https://www.bipm.org/en/committees/jc/jcgm/publications JCGM 100:2008(E)
[26] Meurer A *et al* 2017 SymPy: symbolic computing in Python *PeerJ Comput. Sci.* **3** e103
[27] Virtanen P *et al* 2020 SciPy 1.0: fundamental algorithms for scientific computing in Python *Nat. Methods* **17** 261–72
[28] Harris C R *et al* 2020 Array programming with NumPy *Nature* **585** 357–62

IOP Publishing

Melanie Po-Leen Ooi, Arvind Rajan, Ye Chow Kuang and Serge Demidenko

Chapter 6

Real-world design optimisation problems: applications and usefulness

Moment-based uncertainty propagation methods sidestep the complications of finding the MPP that become apparent in the FORM. Furthermore, the moment approach provides optimiser stability (or convergence) compared to the MC method. Therefore, by utilising the analytical high-order moments accompanied by one of the parametric distribution fitting techniques, an arbitrary system response can be sustained through the use of robust design optimisation (RDO), reliability-based design optimisation (RBDO), or reliability-based robust design optimisation (RBRDO). Such an approach overcomes the limitations of the MPP methods as well as those of other moment-based approaches, thus providing more dependable results in probabilistic design optimisation.

This chapter applies the moment-based uncertainty evaluation method to several real-world examples to show its applicability in generating dependable and sturdy engineering systems.

6.1 The framework for probabilistic design optimisation

Figure 6.1 presents the unified moment-based probabilistic optimisation framework, which will be referred to as the *polynomial moment* (PolyMoment) -based approach in the following discussions. This framework is generic and can be simplified depending on the optimisation paradigm employed, i.e. RDO, RBDO, or RBRDO. The steps used to apply the PolyMoment approach are elaborated in this section, which makes reference to the previous chapters of this book. The remainder of this chapter presents real-world applications of PolyMoment.

As shown in the flowchart, once the probabilistic optimisation problem has been formulated by identifying the objective, performance, and constraint functions according to (3.2), (3.9), or (3.20), the following five steps are executed:

doi:10.1088/978-0-7503-4931-4ch6

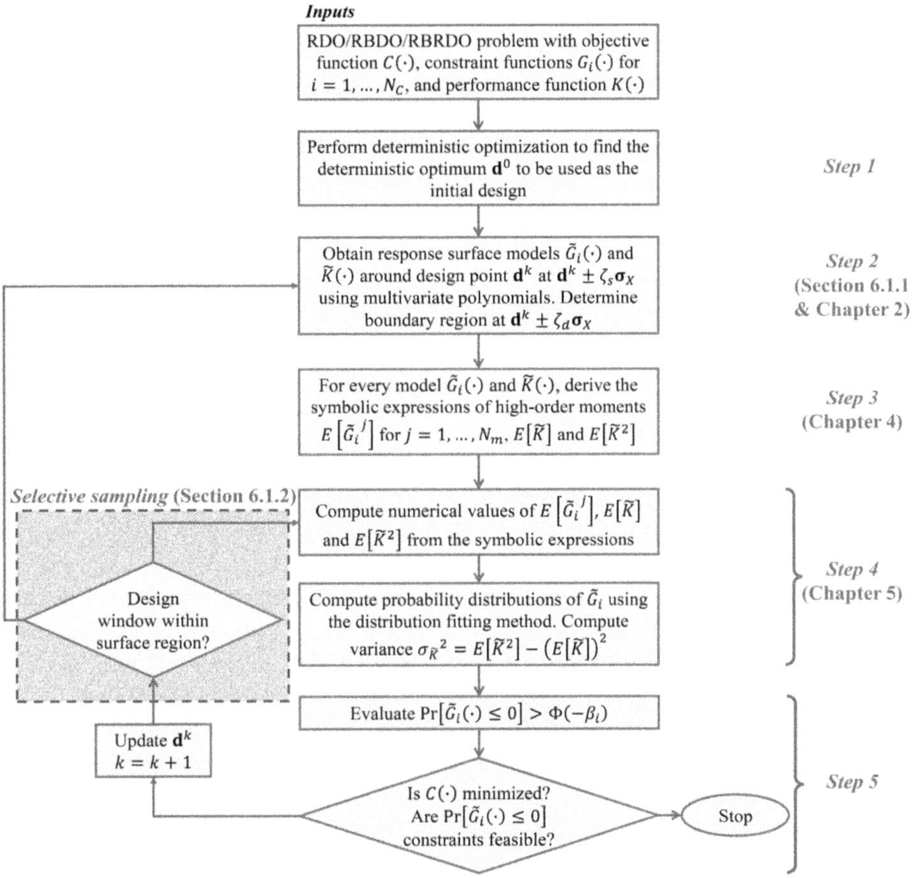

Figure 6.1. The framework of the polynomial moment (PolyMoment)-based probabilistic optimisation for RDO, RBDO, and RBRDO.

Step 1: Find the deterministic optimum design using the formulation shown in (3.1) and use it as the initial design $\mathbf{d}^{k=0}$ of the moment-based probabilistic optimisation; $\mathbf{d}^k$ denotes the design point at the kth iteration.

Step 2: Model the response surface of the constraint function $G_i(\cdot)$ for $i = 1,\ldots,N_c$ (for RBDO/RBRDO) and the performance function $K(\cdot)$ (for RDO/RBRDO) using a polynomial (refer section 6.1) around design point $\mathbf{d}^k$ using the surface region constant ζ_s and the vector of standard deviation of the random variables $\boldsymbol{\sigma}_X$; then determine the *design window* for *selective sampling* using the design window constant ζ_d and $\boldsymbol{\sigma}_X$ (refer to section 6.1.2).

Step 3: For RBDO and RBRDO, find the closed-form expressions of the high-order moments $E\left[\widetilde{G}_i^{\,j}\right]$ for $i = 1,\ldots,N_c$ and $j = 1,\ldots,N_m$ using the analytical moment propagation framework introduced in chapter 4. For RDO and RBRDO, find $E[\widetilde{K}]$ and $E[\widetilde{K}^2]$. Here, $\widetilde{G}_i(\cdot)$ and $\widetilde{K}(\cdot)$ denote the approximated $G_i(\cdot)$ and $K(\cdot)$, respectively. Evaluate the numerical values of the moments.

Step 4: For RBDO and RBRDO, find the probability distribution of the performance functions using one of the moment-based distribution fitting methods studied in chapter 5. For RDO and RBRDO, compute the variance of $\widetilde{K}$ using $\sigma_{\widetilde{K}}{}^2 = E[\widetilde{K}^2] - (E[\widetilde{K}])^2$.

Step 5: Finally, ensure that the constraints $\Pr[G_i(\mathbf{d}, \mathbf{X}) \leqslant 0] > \Phi(\beta_i)$ are met using the estimated probability distribution (for RBDO and RBRDO) and that the objective function $C(\cdot)$ has been minimised. If either of the conditions is not met, update design point $\mathbf{d}^k$ for $k = k + 1$ using the chosen optimiser and go to step 2 if the design window exceeds the response region; or step 3 otherwise (refer to section 6.1.2).

6.1.1 Local response surface modelling using multivariate polynomials

In problems whose large-scale optimisation would involve the use of costly simulations, metamodels (or surrogate models) are used instead of actual simulations to assess the performance and constraints [1]. Polynomial metamodels are a popular choice for this purpose [2, 3]. Studies [2, 4] have shown that the use of polynomial metamodels can significantly reduce the cost of probabilistic optimisation, even when functional relationships between performance functions and input variables are available. Section 3.2.3 covers techniques commonly used to express system responses using multivariate polynomials. The choice of model for the PolyMoment framework given in figure 6.1 is up to the user.

Although polynomial metamodels can greatly improve computational efficiency, using them as a global approximation sacrifices accuracy when the constraint and performance functions are highly nonlinear. Therefore, local approximations of the responses can be made at every iteration. Local approximations using second-order polynomials are some of the most commonly used approaches, even in MPP-based optimisation methods [2, 4]. The results of these studies have demonstrated that this strategy significantly improves both the computational efficiency and the accuracy of the RBDO computation. The next subsection presents a selective sampling technique introduced in [2] for the surface reconstruction procedure, which is a way of further boosting the computational efficiency.

6.1.2 The selective sampling technique

Selective sampling, as proposed by [2], balances global and local modelling by determining the need for a new local response surface based on a mechanism that uses a *surface region* and a *design window* (explained below). This approach improves computational efficiency by only constructing a new local response surface when deemed necessary.

As shown in figure 6.1, the PolyMoment-based optimisation starts by finding the deterministic optimum $\mathbf{d}^{k=0}$; then, the response surface is locally approximated around the intervals:

$$\mathbf{d}^k \pm \zeta_s \boldsymbol{\sigma}_X. \tag{6.1}$$

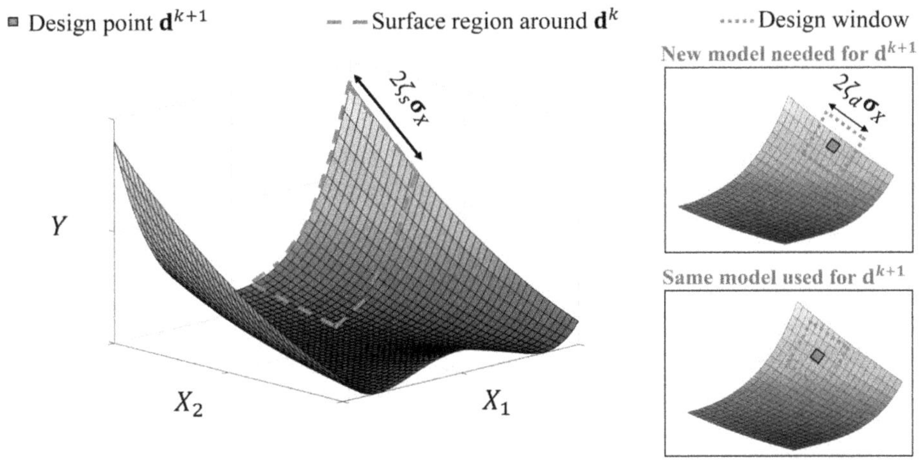

Figure 6.2. An illustration of the selective sampling technique using the highly nonlinear Rosenbrock function [6]. The figure shows the situation in which a new model is required when the design point's design window exceeds the surface region and vice versa.

The value of the surface region constant ζ_s is typically selected based on a multiplicative factor of 1.2–1.5 times the target reliability index β used in the optimisation problem.

For example, 1.2β is used for moderately nonlinear problems and 1.5β is used for highly nonlinear ones [5]. The design window constant ζ_d is a smaller value ($\zeta_d < \zeta_s$) used for the selective sampling technique. The mechanism of the technique is illustrated in figure 6.2 using the highly nonlinear Rosenbrock function [6].

Consider the case in which a local response surface (depicted by dashed lines) is constructed around $\mathbf{d}^k$ using the intervals (6.1). In the next iteration, the design point $\mathbf{d}^{k+1}$ is changed and a decision about the need for a new model is made using the design window, which is defined as $\mathbf{d}^{k+1} \pm \zeta_d\boldsymbol{\sigma}_X$ (depicted by dotted lines). As shown in figure 6.2, when the design window around $\mathbf{d}^{k+1}$ exceeds the boundaries of the existing surface region, a new local response surface is constructed around $\mathbf{d}^{k+1}$. On the other hand, if the design window remains within the existing surface region, the current model is used. As a result of implementing the selective sampling technique, the local response surface does not have to be created at every iteration, thus improving the overall efficiency of the probabilistic optimisation process.

6.2 Lithium-ion batteries: a reliability-based design optimisation framework

In complex engineering systems, the relationship between the output of interest and the input variables is not always clear, so numerical algorithms such as the finite element method are used to calculate the output variables. The input–output relationship in an FE model can only be approximately calculated using basic mathematical models, which are then locally evaluated for further examination including uncertainty evaluation. A prime example is the modelling of the mechanical strength of lithium-ion batteries when they are subjected to sudden impact [7]. The goal is to understand the relationship between the mechanical strength of the battery and three inputs: displacement, surrounding temperature, and strain rate. The model should consider uncontrolled factors, such as manufacturing tolerances, shocks, and vibrations, which affect the battery strength estimation. The purpose of the model is to study the behaviour of the battery's mechanical strength and determine optimal input values that result in the minimum or maximum mechanical force.

Modelling the mechanical strength of a *lithium-ion battery* (LIB) when it is subjected to sudden impact is a complex task due to various variables such as stochastic and dynamic loads, variations in material properties, and interactions among elements. The force developed in an LIB during a crash is an important factor in determining its resistance to tensile impact and serves as a basis for its new mechanical design.

To obtain accurate results, the uncertainties of the inputs and their impact on the mechanical strength must be considered using an uncertainty evaluation toolbox, as discussed in previous chapters of the book. An FE model is used as an initial step to analyse the mechanical forces experienced by the battery in relation to its inputs through the use of surface plots or optimisation algorithms. However, there are two challenges associated with this method. First, repeating the FE model's execution for uncertainty analysis in the generation of surface plots or optimisation significantly increases the computational load. Second, the output mechanical force calculated using the FE model does not account for the uncertainties of the inputs.

A Monte Carlo simulation was considered but discarded, even though it is the most robust and effective method for this type of uncertainty evaluation. MC simulation involves the generation of multiple independent realisations of input quantities using known probability distributions, which are then used to calculate the output probability distribution. The mean and standard deviation of the output can also be obtained. However, the computational time required to run the MC simulation can be prohibitively long if the FE model takes 10 min per execution and 104 realisations are required to obtain a single distribution.

Instead, this case study employs an *artificial neural network* (ANN) to obtain a global model of the battery's mechanical strength and applies the PolyMoment-based approach (see figure 6.3) to reduce the computational load. This approach considers variations in the inputs and provide high-order statistics such as the mean, standard deviation, skewness, and kurtosis, as well as the probability distribution of the mechanical force.

6.2.1 The finite element model of the lithium-ion battery

The ABAQUS/Explicit version 6.14 software [8] is used to model the mechanics of the lithium-ion battery with fully coupled thermal stress analysis to understand the interaction effects of various factors on the maximum crushing load of the battery pack. The model, described in [7], includes components such as the battery casing,

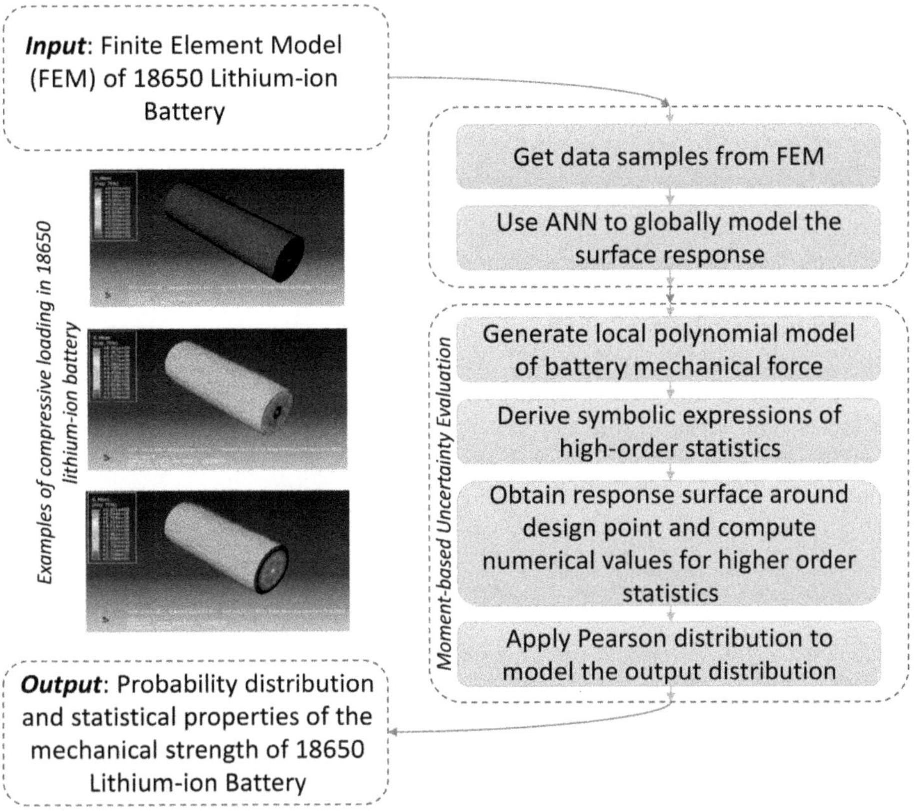

Figure 6.3. A framework for RBDO that utilises moment-based uncertainty evaluation with an FE Model of the 18650 LIB to obtain the probability distribution and statistical properties of its mechanical strength.

jelly roll, composites, and isolators. Smaller components are ignored, as their deformation is largely insignificant according to [9].

It should be noted that studies such as [10–12] have developed constitutive mechanical models for the failure assessment of battery packs. In this study, the battery casing, which is made of steel, and the jelly roll, modelled as crushable foam, undergo significant plastic deformation during mechanical loading and are considered in the analysis. This study uses a homogenised mechanical model of the jelly roll from [10] to reduce computational time, while still accurately replicating the load and displacement of the 18650 LIB. This allows for finer meshing and more accurate deformation and failure mechanism predictions in response to the applied load.

The FE model was validated by comparing simulation results with experimental data, which showed good agreement. The force–displacement plot, force–temperature plot, and force–strain rate plot were compared for the results of the simulation and the experiment; the comparison revealed similar levels of accuracy. The FE model was used to obtain compressive force data for variations in displacement, strain rate, and temperature. The FE model snapshots (see examples on the left-hand side of figure 6.3) showed the effect of compression loading on the battery structure.

6.2.2 Incorporating moment-based uncertainty evaluation

In this study, data samples were collected from the finite element model using *Latin hypercube sampling* [13] and an ANN was trained using 70% of the collected data to build a global model of the mechanical force in a lithium-ion battery, as shown in [7]. The temperature values ranged from 10 °C to 50 °C with an interval of 10 °C, the displacement ranged from 2mm to 8mm with an interval of 2mm, and the strain rate ranged from 0.05 to 0.20 mm s^{-1} with an interval of 0.05 mm s^{-1}. The ANN model showed high prediction accuracy with a coefficient of determination (R^2) of 0.999 74 (figure 6.4).

To consider the uncertainties in the inputs and obtain the statistical properties and probability distribution of the mechanical force, the model-based uncertainty evaluation described in section 4.2 was performed on the ANN model. This evaluation required the use of a multivariate polynomial to represent the relationship between the inputs and the outputs, and therefore a local model of the mechanical force F_l was established using a general three-variable second-order polynomial equation:

$$F_l(X) = a_{00} + \sum_{i=1}^{N} a_{0i}X_i + \sum_{i=1}^{N} a_{ii}X_i^2 + \sum_{i=1}^{N}\sum_{j=i+1}^{N} a_{ij}X_iX_j + \epsilon, \tag{6.2}$$

where $\boldsymbol{X} = \{X_1,\ldots,X_N\}$ denotes the input variables, $N = 3$ is the number of variables, $\{a_{00}, a_{0i}, a_{ii}, a_{ij}\}$ is the set of model coefficients, and ϵ is the modelling (or approximation) error.

The symbolic expressions for the higher-order statistics of the mechanical force were then computed using a toolbox described in section 4.2. It is important to note that only the finite element runs used to train the ANN model were needed to compute the probability distribution of the mechanical force. No additional finite element simulations were required, as the changes in the response surface model

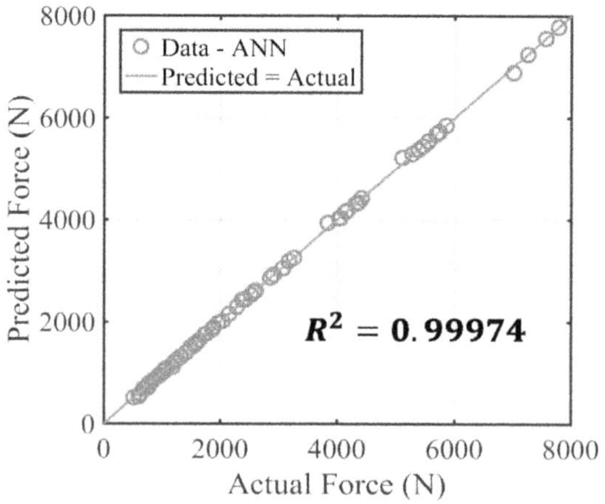

Figure 6.4. A comparison between the mechanical force predicted by the ANN model and the actual experimental data obtained from the finite element (FE) model. Reprinted from [7], Copyright 2018, with permission from Elsevier.

were captured in the coefficients. However, the form of the model remained unchanged.

Next, the coefficients of the polynomial model were obtained using the Box–Behnken design approach for the design of experiments and the least-squares method [14]. The data samples were collected for the design of experiments to obtain the numerical values of the higher-order statistics. The local model was constructed within a range of $\pm 3\sigma X$, where σX is the standard deviation vector of the input variables.

The least-squares method was employed to build the model, making it possible to assume that the error term ϵ followed a normal distribution with zero mean and standard deviation $\sigma\epsilon$ [14]. This additional random variable was incorporated when using the toolbox from section 4.2 to compute the higher-order statistics, considering the modelling error of F_l when calculating its uncertainty. The calculated higher-order statistics were then used with the Pearson system (described in section 2.5.5) to approximate the distribution of the mechanical force.

6.2.3 The resultant design

The probability distributions obtained through the PolyMoment-based RBDO method are compared to those obtained through Monte Carlo simulation using finite element simulation data in figure 6.5. The comparison is made for a displacement of 5mm, a temperature of 30 °C, a strain rate of 0.125 mm s^{-1}, and with coefficients of variation of 0.01 and 0.05. The histogram of the Monte Carlo

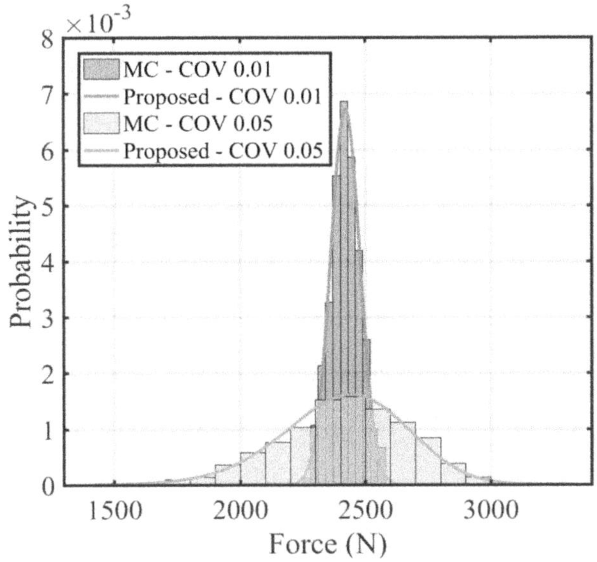

Figure 6.5. A comparison of the probability distributions produced by PolyMoment-based RBDO and a Monte Carlo simulation using the FE data for a displacement of 5mm, a temperature of 30 °C, and a strain rate of 0.125 mm s^{-1} with coefficients of variance (COVs) of 0.01 and 0.05. Reprinted from [7], Copyright 2018, with permission from Elsevier.

simulation was generated with a sample size of 500. The results in figure 6.5 demonstrate that the distributions obtained from the proposed framework are in good agreement with the Monte Carlo simulation. In addition, the figure shows that as the coefficient of variation of the inputs increases, the spread of the mechanical strength distribution also increases. Although the study assumes normal distributions for the inputs, the proposed method is capable of handling non-normal distributions as well.

PolyMoment is capable of performing probabilistic analyses under various uncertainty conditions, including different distributions and input uncertainties, without requiring additional ANN model building or FE simulation evaluations. This capability is of significant value, especially when FE models require high computational times for each execution or when design optimisation algorithms necessitate repeated evaluations of output probability distributions or sensitivities.

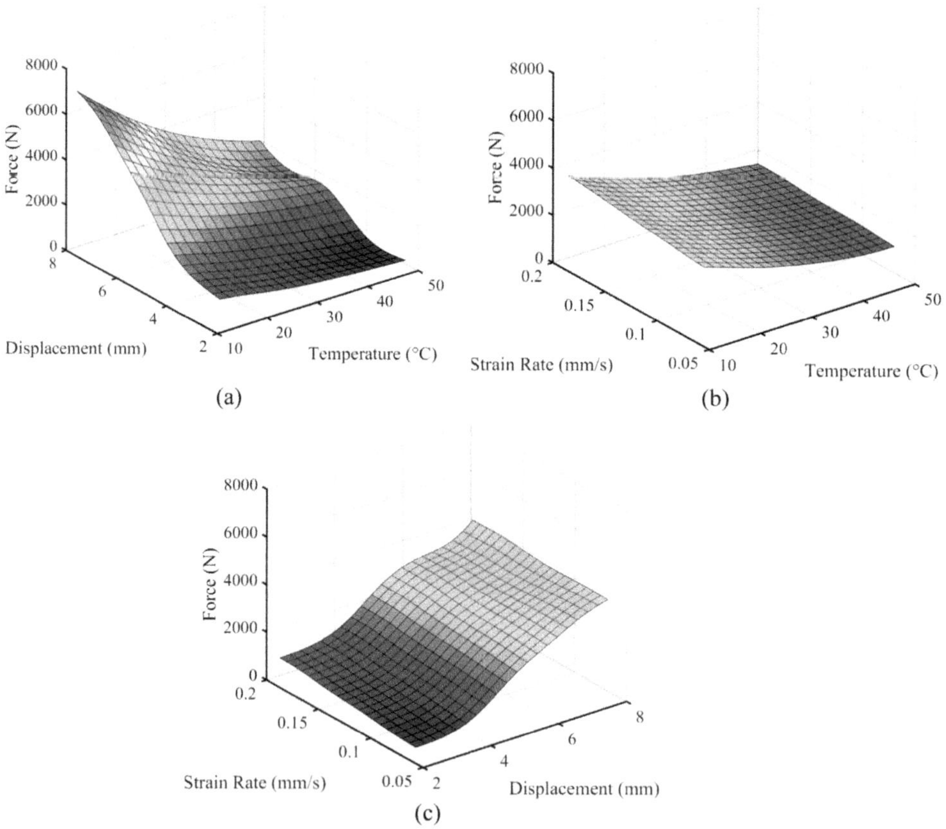

Figure 6.6. A depiction of the relationship between the input variables and the mean mechanical strength of an LIB; one input is held constant in each plot. (a) A constant strain rate of 0.19 mm s^{-1}, (b) a constant displacement of 7.66 mm, and (c) a constant temperature of 10.47 °C. The input variables are assumed to follow a normal distribution with a coefficient of variation of 0.01. Reprinted from [7], Copyright 2018, with permission from Elsevier.

The mean mechanical force computed by the PolyMoment framework was used to generate the three surface plots in figure 6.6, which illustrate the influence of inputs on the mechanical force experienced by the battery. By keeping one of the inputs constant, it is possible to observe the trend in figure 6.6. This trend remains consistent regardless of the constant values used, allowing one to infer the sensitivity of the force experienced by the battery force with respect to its inputs.

The surface plots in figure 6.6 demonstrate that the mechanical strength of the battery increases with increasing strain rate and displacement, but decreases with increasing temperature. These results are aligned with experimental results, which show that larger external impacts result in higher mechanical stresses and strains. Higher temperatures decrease battery strength due to decreased hardness and toughness. The sequential quadratic programming algorithm [15] and other advanced optimisation methods (genetic algorithm [16], stepwise optimisation [17, 18]) show that the maximum mean mechanical strength (7530 N) can be achieved at displacement of 8 mm, a temperature of 10 °C, and a strain rate of 0.2 mm s^{-1}.

From the plots, it can be concluded that: (1) the proposed probabilistic framework accurately predicts the probability distribution of the LIB's mechanical strength; (2) lower displacement and temperature result in higher battery strength; (3) displacement and temperature have equal impacts on battery strength, while the strain rate has a lower impact.

The application of the PolyMoment framework offers a probabilistic method for battery design that takes account of uncertainties both inside and outside the battery, which is vital for battery dependability and security. These findings are especially relevant to battery manufacturers who aim to meet safety regulations and prevent accidents in electric vehicles.

6.3 Vehicle design based on side-impact crashworthiness: the application of a reliability-based robust design optimisation problem

This section examines a vehicle side-impact crashworthiness study [19], which is an optimisation problem with nine design variables, eleven random variables, and ten constraint functions and is commonly used as a benchmark for new probabilistic optimisation methods. The aim is to reduce the weight of the vehicle $W(\mathbf{X})$ and the variance of the performance function $K(\mathbf{X})$ while improving side-impact crash protection for passenger safety. To achieve this, the European Enhanced Vehicle-Safety Committee side-impact procedure [20] is utilised to establish the reliability constraints of the performance functions.

6.3.1 Problem formulation

The problem formulation (from [19]) is:

$$
\begin{aligned}
\text{minimize: } & C = (\mathbf{X}) = w_1 \frac{W}{W_0} + w_1 \frac{\sigma_K^2}{\sigma_0^2}, \\
\text{subject to: } & \text{abdomenload} < 1\text{kN}, \\
& \text{upper/middle/lower viscous criteria} < 0.32 \ \text{ms}^{-1}, \\
& \text{upper/middle/lower rib deflection} < 32 \ \text{mm}, \\
& \text{pubicsymphysisforce} < 4\text{kN}, \\
& \text{velocity of B–pillar at middle point} < 9.9 \ \text{mm ms}^{-1}, \\
& \text{velocity of front door at B–pillar} < 15.7 \ \text{mm ms}^{-1}, \\
\text{where: } & \mathbf{d}^L \leqslant \mathbf{d} \leqslant \mathbf{d}^U, \ \ \text{and } \beta = 1.282,
\end{aligned}
\tag{6.3}
$$

where $w_1 = 0.5$ and $w_2 = 0.5$ are the weights, $W_0 = 29.05$ kN is the initial weight, and $K_0 = 1.4781$ mm is the initial variance of the performance function based on the original design points given in [19]. The constraint and performance functions, initial design points, statistical information, and details of the random variables are given in [19]. The target reliability index for all constraint functions is $\beta = 1.282$.

6.3.2 Resultant design

The performance function that represents the safety component of the design objective in this RBRDO problem is the lower rib deflection. To assess its robustness and reliability, we adopt a combination of moment-based robustness analysis using univariate DRM and PMI (from section 3.1) and reliability analysis using PMA (from section 3.2). The results are compared with those obtained from the PolyMoment-based RBRDO method, as depicted in figure 6.1, which provides a comprehensive framework for performing both probabilistic analyses. The final design outputs are tabulated in table 6.1.

In addition to the total number of function evaluations and the accuracy of meeting the probability constraints, the accuracy with which $\sigma_K$is computed is also considered as a performance metric. Unlike the previous examples, the actual reliabilities computed by the MC method are presented in a radar chart in figure 6.7. Since the failure probability is 10% ($\beta = 1.282$), the ideal solution is a radar line that stays within the two outermost polygon regions. Any radar line that crosses the

Table 6.1. The performance of the robustness analysis of the PMI, DRM, and proposed PolyMoment methods in vehicle side-impact crashworthiness evaluation. FEV denotes the function evaluation counts [21].

	Optimum design			
Method	W (kN)	Calculated σ_K (mm^2)	Actual σ_K (mm^2)	Total FEV counts
PMI and PMA	28.7780	0.0618	0.6254	4590
DRM and PMA	25.8685	0.4144	0.4645	4940
PolyMoment	27.6034	0.1633	0.1635	1526

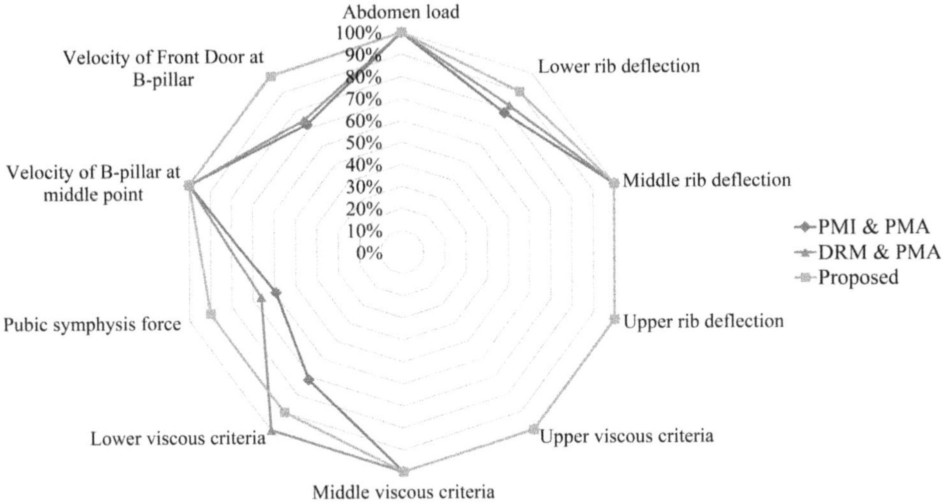

Figure 6.7. The probability of satisfying the constraint functions for the vehicle side-impact crashworthiness problem. Reproduced from [21], CC BY 3.0.

second outermost boundary towards the centre indicates a design violation for that specific reliability constraint.

Table 6.1 compares the results of three RBRDO techniques in terms of side-impact crashworthiness. All three methods offer a significant reduction in vehicle weight and lower rib cage deflection variance. However, the PolyMoment method outperforms the others by yielding optimal results three times faster.

The PolyMoment method also produces a design that offers a good balance of vehicle weight and safety, while the variance estimations produced by the MC method reveal that the DRM and PMI methods underestimate the variance. This results in lighter but less safe designs. For example, if the DRM is used, the variance is underestimated at 0.4144 mm^2, which might be overlooked by the designer, as the true value is 0.4645 mm^2. Increasing the quadrature points or using bivariate DRM could improve accuracy, but at the cost of increased function evaluations. The PMI method has an even greater variance underestimation.

Although the PMA with DRM appears to be the best option for weight optimisation based on table 6.1, the radar chart in figure 6.7 indicates that this design is unreliable and fails to meet safety standards when validated using the MC method. This aligns with the literature [22], which states that multiple MPP iterations can lead to unreliable designs. However, the design solution produced by the PolyMoment method always satisfies the failure probability constraint of less than 10%.

6.4 Fuel cells: parameter optimisation for reliable and robust operation

Hydrogen fuel cells are seen as a promising energy storage technology due to their environmental friendliness and efficiency compared to traditional combustion technologies. The polymer electrolyte membrane (PEM) fuel cell is considered one of the most promising types due to its high power output, low operating temperature, efficiency, high current density, and structural safety [23–26]. These benefits make PEM fuel cells suitable for small-scale power generation and for use in the automobile industry.

The operation of a PEM fuel cell is shown in figure 6.8. It consists of an electrolyte sandwiched between two electrodes (the anode and cathode), with bipolar plates on both sides that distribute hydrogen and oxygen gases and serve as current collectors. Hydrogen gas flows to the anode, where it is separated into protons and electrons by a catalyst; the protons flow through the membrane to the cathode and the electrons flow through an external circuit to generate electricity. On the cathode side, oxygen reacts with the hydrogen ions to form water, and the exothermic reaction generates heat [27].

Research has been conducted on PEM fuel cells to optimise their efficiency. The first fuel cell models date back to the early 1990s [27]. More comprehensive models were later developed, including a dynamic model that incorporated six submodels [29]. Other studies optimised operational parameters using novel algorithms [30, 31], a battery–capacitor hybrid system [32], and stochastic dynamic programming [33]. However, no study has considered the uncertainties in the operating parameters and their propagation to output uncertainties in PEM fuel cell optimisation.

This subsection utilises a PolyMoment framework that incorporates uncertainty evaluation into the design optimisation loop of a PEM fuel cell. This approach will result in a more robust and cost-efficient PEM fuel cell, taking into account factors

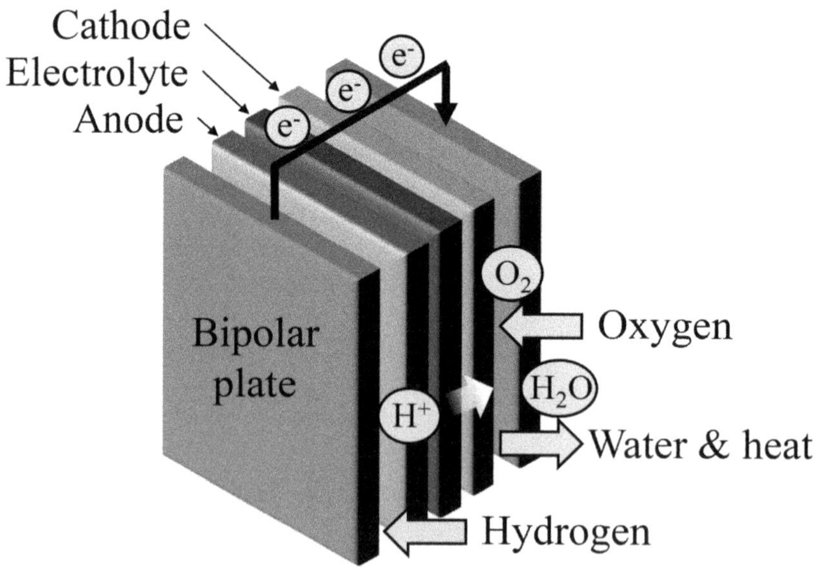

Figure 6.8. The polymer electrolyte membrane (PEM) fuel cell. Reprinted from [28], Copyright 2018, with permission from Elsevier.

such as manufacturing variations, material variations, and uncontrollable operating environments [34–38].

6.4.1 Problem formulation

The problem statement focuses on optimising the output power of a PEM fuel cell described in [29] while considering the hydrogen mass flow rate (equation (6.4)) in an economical design. The study uses a dynamic fuel cell model made up of six components and considers the hydrogen mass flow rate to be an important parameter. The output power of the fuel cell is dependent on five inputs, and the study aims to analyse the uncertainty of the outputs (fuel cell power and hydrogen flow rate) and perform a sensitivity analysis to identify the optimal parameters for a reliable and robust design. The PolyMoment framework is proposed to perform the analysis in an accurate and efficient manner.

$$W_{\mathrm{H_2}} = \frac{I_{\mathrm{st}}}{2F} M_{\mathrm{H_2}} n \lambda_{\mathrm{an}}, \tag{6.4}$$

where I_{st} is the stack current, F is the Faraday constant, $M_{\mathrm{H_2}}$ is the molecular mass of hydrogen, n is the number of cells, and λ_{an} is the hydrogen excess rate.

The efficiency of the fuel cell's output power is influenced by various parameters, such as the stack current I_{st}, the stack temperature T, the oxygen excess ratio λ_{ca}, the hydrogen excess ratio λ_{an},, and the inlet air humidity $\varphi_{\mathrm{ca,\ in}}$ (refer to figure 6.9). These operating conditions are usually optimised for maximum output power efficiency and minimum hydrogen consumption. To assist this optimisation process, response surface methodology (RSM) is used to describe the relationship between the output and the inputs of the system. This study takes account of the uncertainties in fuel cell output power and hydrogen consumption in the optimisation process, as illustrated in figure 6.10.

The response surface of the PEM fuel cell's output power is constructed using a second-order polynomial model:

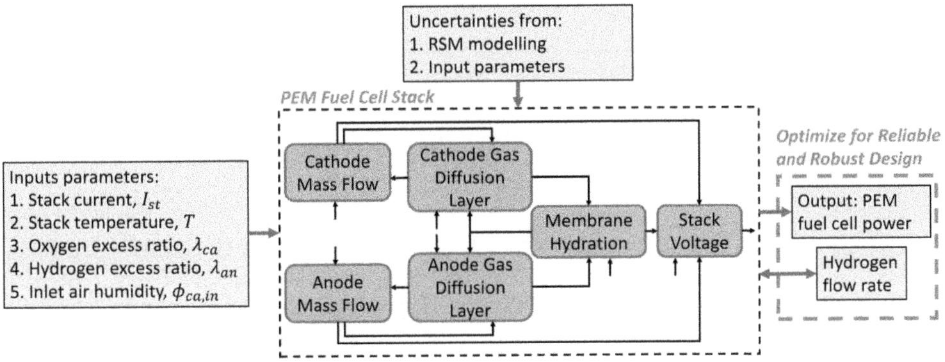

Figure 6.9. The relationships and modelling of variables in the optimisation of PEM fuel cell design with a focus on optimising power subject to a low hydrogen mass flow rate. Reprinted from [28], Copyright 2018, with permission from Elsevier.

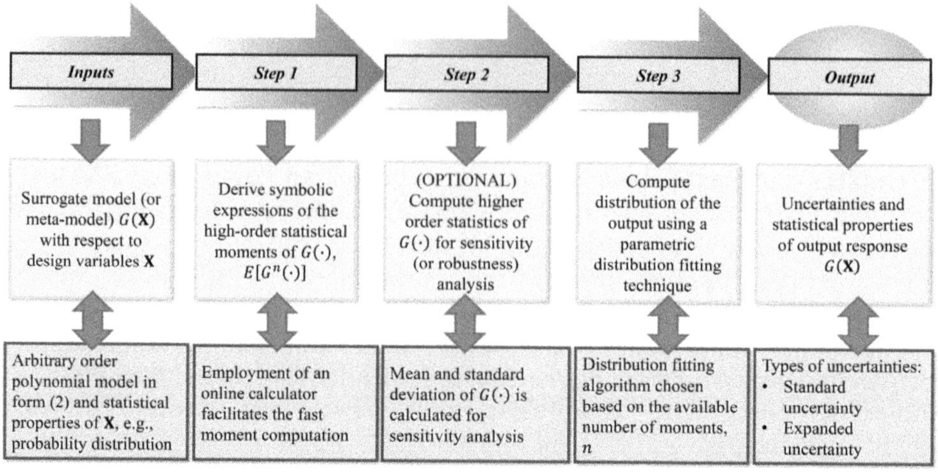

Figure 6.10. The framework used to apply PolyMoment to determine the statistical properties of the PEM fuel cell. Reprinted from [28], Copyright 2018, with permission from Elsevier.

Table 6.2. The minimum and maximum values of the design variables used to construct the RSM model of PEM fuel cell output power and the statistical properties (probability distribution and coefficient of variation) attributed to the respective design variables for probabilistic analysis.

Design variable	Minimum value	Median value	Maximum value	Probability distribution	Coefficient of variation
Stack current, I_{st} (A)	36.0	43.0	50.0	Uniform	0.01
Stack temperature, T (°C)	55.0	60.0	65.0	Uniform	0.01
Oxygen excess ratio, λ_{ca}	1.5	2.5	3.5	Normal	0.01
Hydrogen excess ratio, λ_{an}	1.1	1.3	1.5	Normal	0.01
Inlet air humidity, $\phi_{ca,\,in}$	0.6	0.8	1.0	Uniform	0.01

$$G(\mathbf{X}) = a_{00} + \sum_{i=1}^{N} a_{0i}X_i + \sum_{i=1}^{N} a_{ii}X_i^2 + \sum_{i=1}^{N-1}\sum_{j=i+1}^{N} a_{ij}X_iX_j + \varepsilon. \tag{6.5}$$

Here, $G(\cdot)$ denotes the output response, $\mathbf{X}$ denotes the vector of design variables, N denotes the number of design variables, and $\{a_{00},\ a_{0i},\ a_{ii},\ a_{ij}\}$ is the set of model coefficients. Table 6.2 presents the lower and upper bounds of the input stack current, stack temperature, oxygen excess ratio, hydrogen excess ratio, and inlet air humidity.

The first step is to derive symbolic expressions for the high-order moment of $G(\cdot)$ using the toolbox in section 4.2. The next step involves substituting the numerical values of the model coefficients and the other parameters of X to obtain the statistical parameters of $G(\cdot)$. The optional third step is a sensitivity or robustness analysis.

6.4.2 Sensitivity analysis

This study aims to find the optimal combination of design variables that results in the highest output power P and the lowest hydrogen mass flow rate W_{H_2} for a PEM fuel cell. To ensure economical operation, it is crucial to minimise W_{H_2}. To determine the design's robustness, the sensitivities of the mean and standard deviation of P and W_{H_2} with respect to the stack current I_{st} and the hydrogen excess ratio λ_{an} are analysed using the 3D surface plot shown in figure 6.11.

The other parameters were kept constant and were assumed to have a coefficient of variation of 0.01. As can be seen in figures 6.11(a) and (b), the mean and standard deviation of the output power are more sensitive to changes in I_{st} than to changes in λ_{an}. However, both I_{st} and λ_{an} have an equal impact on the mean and standard deviation of W_{H_2}, as shown in figures 6.11(c) and (d).

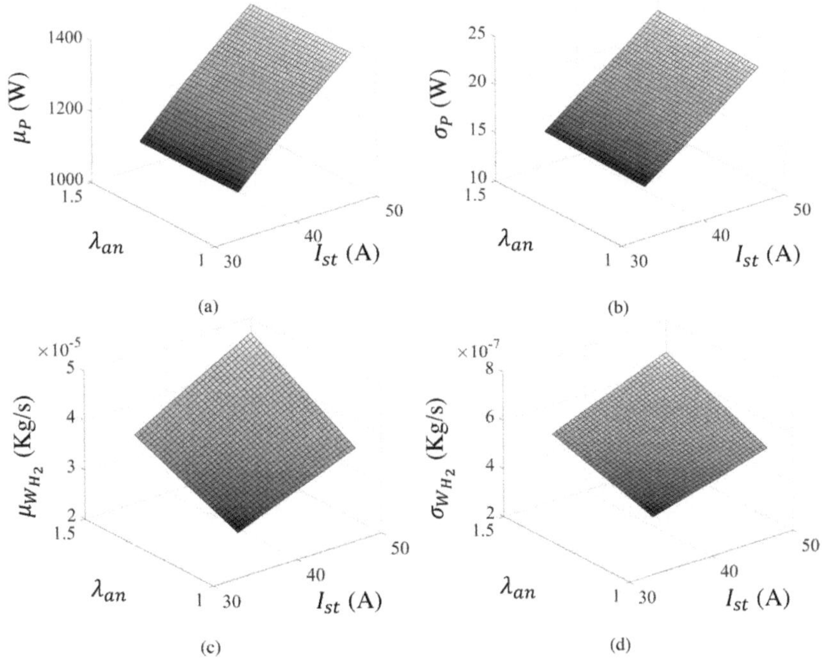

Figure 6.11. Surface plots showing the influence of the stack current I_{st} and hydrogen excess ratio λ_{an} on the (a) mean and (b) standard deviation of the output power P as well as (c) on the mean and (d) standard deviation of the hydrogen mass flow rate. The other parameters are kept constant, as follows: the stack temperature is 60 °C, the oxygen excess ratio is 2.5, and the air humidity ratio is 0.8. All the parameters are assumed to have distinct probability distributions with a coefficient of variation of 0.01. Reprinted from [28], Copyright 2018, with permission from Elsevier.

6.4.3 Determination of the optimal operating conditions for RBDO and RBRDO

There are two ways to optimise the design of a PEM fuel cell for optimal power output while maintaining a low hydrogen flow rate of 3e−5 kg s^{-1}: (1) reliability-based optimisation with a 99% confidence constraint and (2) reliability-based robust optimisation, which also minimises the standard deviation of the output power and the hydrogen flow rate.

The objectives and constraints are set using a sequential quadratic programming algorithm [39], and the PolyMoment framework is used to calculate the reliability and the standard deviation. The Pearson system of distributions is used for approximation. The output power is represented by equation (6.5) with $N = 5$, and the hydrogen mass flow rate is given by equation (6.4).

For this case study, the problem of reliability-based optimisation can be mathematically formulated as:

$$\begin{aligned} &\text{Maximise: } \mu_{G(\mathbf{d},\,\mathbf{X})},\\ &\text{subject to: } W_{H_2}(\mathbf{d},\,\mathbf{X}) \leqslant 3e^{-5}\ \text{kg},\\ &\quad\text{where: } \mathbf{d}^{L} \leqslant \mathbf{d} \leqslant \mathbf{d}^{U} \text{ and } \mathbf{d} = E[\mathbf{X}], \end{aligned} \tag{6.6}$$

where $\mathbf{d}^L$ and $\mathbf{d}^U$ are the lower and upper bounds of the vector of design variables $\mathbf{d}$, respectively.

Based on the optimisation results, the highest mean output power of 1329.56 W can be achieved with the following parameters: a stack current of 42.16 A, a stack temperature of 63.98 °C, an oxygen excess ratio of 3.50, a hydrogen excess ratio of 1.10, and an inlet relative air humidity of 0.63. These values also ensure, with 99% confidence, that the hydrogen flow rate stays below 3e−5 kg s^{-1}. The results are even consistent with those of other advanced optimisation algorithms, such as the genetic algorithm [16] and stepwise optimisation [17, 18].

Figure 6.12 compares the results of the PolyMoment framework with Monte Carlo simulations that include and exclude uncertainties in the hydrogen mass flow rate. The Monte Carlo simulations, which are considered to represent the current state of the art in the field but are computationally intensive, demonstrate close agreement with the results from PolyMoment. Furthermore, PolyMoment offers a more computationally efficient approach than the Monte Carlo simulations.

However, the Monte Carlo simulation that excludes uncertainties in the hydrogen mass flow rate has a 50% chance of exceeding the design constraint for the hydrogen mass flow rate, which could lead to increased hydrogen consumption and cost. This highlights the importance of considering uncertainties when optimising for the power output in this field. The maximum power output that can be attained without considering uncertainties is 1545.25 W, as reported in [29].

The standard deviations of the output power and the hydrogen flow rate of the PEM fuel cell can be reduced to improve the quality of the design. This results in a robust design that is less affected by external factors and more consistent in its output. PolyMoment can be used to calculate the standard deviation and minimise it

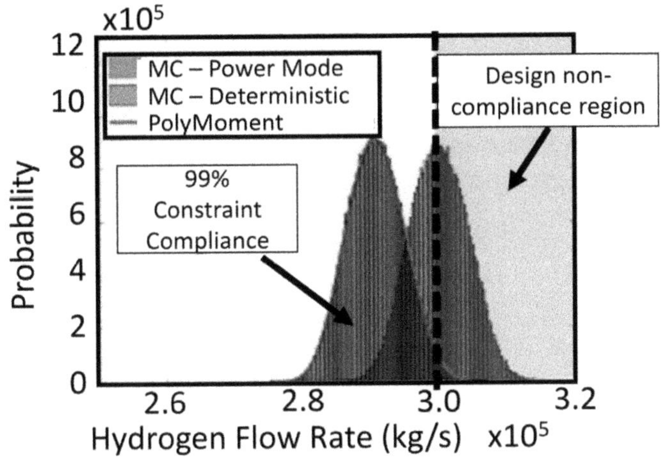

Figure 6.12. A comparison between the probability distribution obtained from the PolyMoment framework and those obtained using Monte Carlo simulations for the optimal parameter values that lead to the maximum output power. Reprinted from [28], Copyright 2018, with permission from Elsevier.

as an additional objective for a robust fuel cell design. Mathematically, the minimisation of these additional parameters can be added into the optimisation algorithm as follows:

$$\begin{aligned} &\text{minimize: } w_1\frac{\mu_G}{\mu_{G^0}} + w_2\frac{\sigma_G}{\sigma_{G^0}} + w_3\frac{\sigma_{W_{H_2}}}{\sigma_{W_{H_2}^0}}, \\ &\text{subject to: } W_{H_2}(\mathbf{d}, \mathbf{X}) \leqslant 3e^{-5} \text{ kg}, \\ &\quad \text{where: } \mathbf{d}^L \leqslant \mathbf{d} \leqslant \mathbf{d}^U \text{ and } \mathbf{d} = E[\mathbf{X}], \end{aligned} \tag{6.7}$$

where G^0 and $W_{H_2}^0$ denote the initial values of the output power and hydrogen flow rate, respectively, and $\{w_1, w_2, w_3\}$ are weights assigned by the design engineer based on design priorities such that $w_1 + w_2 + w_3 = 1$. The standard deviations of both $G(\cdot)$ and $W_{H_2}(\cdot)$ can be symbolically obtained from PolyMoment in figure 6.10.

The impacts of setting different priorities in the optimisation problem (6.7) were studied using three sets of weights $\{w_1, w_2, w_3\}$. Table 6.3 shows the values of three parameters of interest for all three scenarios obtained through MC simulation. The results show that the three scenarios lead to different fuel cell designs. The designer can choose the weights in problem (6.7) to balance the trade-off between maximising the mean power, the robustness of power, and the hydrogen flow rate. A trade-off and cost–benefit analysis is needed for the final optimal operating parameters. PolyMoment is a preferable option for uncertainty evaluation, as it is accurate, computationally efficient, and straightforward to use for design optimisation regardless of the problem's nonlinearity and dimensionality. Incorporating system uncertainties is crucial for a dependable and high-quality PEM fuel cell design, making it a valuable tool for design engineers.

Table 6.3. The differences in the mean and the standard deviation of a fuel cell's output power and the standard deviation of hydrogen flow rate for different weights in the optimisation problem shown in (6.7). The results were obtained using MC simulation.

Weights $\{w_1, w_2, w_3\}$	Mean output power, μ_G (W)	Standard deviation of the output power, σ_G (kg s^{-1})	Standard deviation of the hydrogen flow rate, $\sigma_{W_{H_2}}$ (kg s^{-1})
{1,0,0}	1329.56	12.32	4.11e−7
$\left\{\frac{1}{3}, \frac{1}{3}, \frac{1}{3}\right\}$	1121.32	9.93	3.51e−7
$\left\{\frac{5}{9}, \frac{2}{9}, \frac{2}{9}\right\}$	1328.28	12.20	4.10e−7

6.5 Magnetic sensor module design

The design of magnetic sensors often involves the use of computer simulations, such as the FE modeling, followed by a design optimisation process to determine the optimal parameters that meet system constraints. However, this approach can be computationally expensive for high-dimensional systems and does not account for variations in manufacture that can lead to the production of noncompliant products. To address these shortcomings, this case study presents the use of a combined approach that uses the *parametric model order reduction* (PMOR) method and the reliability-based design optimisation method within the PolyMoment framework.

Specifically, it focuses on the design of a magnetic sensor for a linear motor mover position detector and uses a three-dimensional (3D) model. A new approach is proposed, incorporating *proper orthogonal decomposition* with *dynamic mode decomposition* (POD-DMD) [40, 41], multiparameter moment matching, and a response surface moment-based RBDO method that uses the PolyMoment framework to achieve an accurate and efficient analysis. This method reduces the risk of noncompliance caused by manufacturing uncertainties and provides a faster computational process. The design obtained from the presented method is compared to that obtained from a deterministic optimisation method to demonstrate its effectiveness.

Figure 6.13(a) displays a 3D view of the magnetic sensor and its placement on the stator of the linear motor in the *xyz* plane. The magnetic sensor module includes a Nd–Fe–B *permanent magnet* (PM) and three iron cores, labelled I, II, and III. A Hall integrated circuit is positioned between cores I and II. The PM is magnetised in the negative Y direction and the desired magnetic flux density B_X is measured at the midpoint of the airgap g_2 in the X direction.

The sensor operates by detecting changes in flux density caused by changes in reluctance. Due to the alternating tooth–slot structure of the linear motor stator, the flux density distribution at the midpoint of g_2 is sinusoidal for a two-pole pitch displacement. For the Hall integrated circuit to output a minimum of 1 V peak to peak, the sensor's *peak flux density* (PFD) must be at least 0.1 T. The *total harmonic distortion* (THD) should be less than 1%. To meet these requirements, the sensor must undergo a thorough parametric study and optimisation.

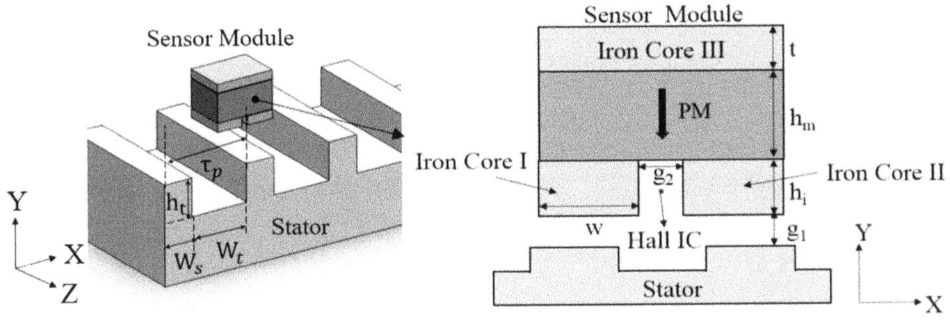

Figure 6.13. A linear position sensor module located on a stator. (a) The arrangement of the sensor module arrangement on the stator and (b) the parameters of the sensor module. PM denotes permanent magnet [42].

6.5.1 Problem formulation

Figure 6.14 shows the PMOR-RBDO framework, which starts with a POD-DMD-based PMOR method to achieve a computationally efficient result and then uses an RBDO technique to find the optimal sensor design that takes account of manufacturing uncertainties. To find the global optimum, this study uses the *genetic algorithm* (GA) as a heuristic method instead of gradient-based optimisers.

Equation (6.8) represents the mathematical model of the magnetic sensor module shown in figure 6.13, which is described as a nonlinear magnetostatic problem with no external electric source. In this equation, $\boldsymbol{M}$ and $\boldsymbol{K}$ are square matrices of size $n \times n$, $\boldsymbol{Y}(t)$ is the state vector, $\boldsymbol{F}(t)$ is the source vector, and $\boldsymbol{N}(t)$ is the nonlinearity associated with the system. The equation shows how the magnetic field $\boldsymbol{MY}(t)$ in the module is affected by the derivative of $\boldsymbol{Y}(t)$, the source vector $\boldsymbol{F}(t)$, and the nonlinearity $\boldsymbol{N}(t)$.

$$\boldsymbol{MY}(t) + \boldsymbol{K}\frac{d\boldsymbol{Y}(t)}{dt} = \boldsymbol{F}(t) + \boldsymbol{N}(\boldsymbol{Y}(t)) \tag{6.8}$$

To make this high-dimensional system more manageable, equation (6.6.1) is transformed into equation (6.6.2) using a combination of the singular value decomposition (SVD)-based POD method and the DMD [40, 43].

$$\boldsymbol{M}_r\boldsymbol{Y}_r(t) + \boldsymbol{K}_r\frac{d\boldsymbol{Y}_r(t)}{dt} = \Phi^*\boldsymbol{F}(t) + \Phi^*\Phi^{\mathrm{DMD}}\mathrm{diag}(e^{w^{\mathrm{DMD}}t})p, \tag{6.9}$$

where $p = (\Phi^{\mathrm{DMD}})^{\dagger}F_1$, $\boldsymbol{M}_r = \Phi^*M\Phi$, and $\boldsymbol{K}_r = \Phi^*K\Phi$. Φ^{DMD} are the DMD bases for rank k, p is the initial condition, and w_i are the eigenvalues. To include the effects of parameters on the system, multiparameter moment matching with the Taylor series is used to parameterise the equation. The magnetic sensor design is optimised using two different scenarios:

1. *Deterministic optimisation*: in this scenario, only the PMOR model is used, which does not account for the uncertainties in the design parameters due to manufacture. This method aims to optimise the peak flux density and total harmonic distortion without considering their tolerance effects.

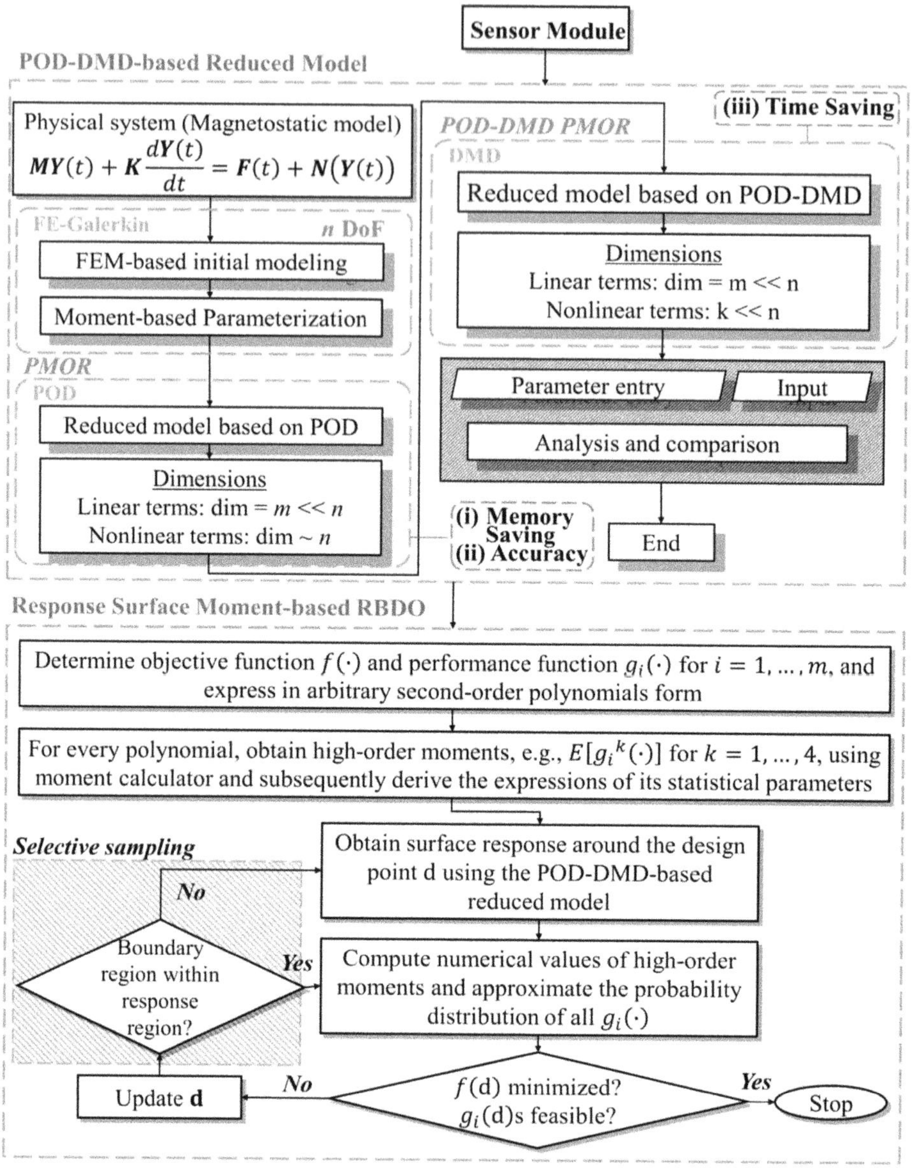

Figure 6.14. The flowchart of the PMOR-RBDO Framework. The RBDO portion of the flowchart, shown in the left half, incorporates the PolyMoment framework [42].

2. *Response surface moment-based RBDO*: in this scenario, the tolerance effects of the design parameters are considered during the optimisation, with a near-negligible failure probability of 0.01%. The optimisation takes account of the uncertainties caused by machines with tolerances of ±0.05 and ±0.1 mm. The actual constraint noncompliance is then calculated using the Monte Carlo method with 10^6 samples.

Table 6.4. Optimal design of the magnetic sensor and constraint compliance for deterministic optimisation and RBDO. FEV denotes function evaluation counts.

		Optimal design		Failure probability			
Optimisation method	Tolerance (mm)	PFD (T)	THD (%)	PFD (%)	THD (%)	Required FEV by GA	PMOR model calls
Deterministic	±0.05	0.13	0.24	0.63	0.00	2200	2200
(Scenario 1)	±0.1	0.13	0.24	10.97	0.00	2200	2200
RBDO (Scenario 2)	±0.05	0.16	0.24	0.00	0.00	58 600	7020
	±0.1	0.20	0.24	0.00	0.00	58 600	6560

6.5.2 The results of the PolyMoment-based RBDO method

The results are summarised in table 6.4, which shows that the deterministic design meets the set constraints but has a higher probability of failing to meet the required PFD constraint than the desired 0.01% because it does not consider the effects of manufacturing tolerance. Figure 6.15 shows that up to 11% of the probability distribution of the PFD could fail to meet the desired constraints for higher manufacturing uncertainties. On the other hand, RBDO avoids this problem and does not compromise the THD, as shown in table 6.4. The results also show that using the PMOR model and the proposed PolyMoment-based RBDO method improves the computational speed significantly compared to using the full FE model. The proposed method provides a significant improvement in terms of design noncompliance cost, computational speed, and accuracy, making it valuable for the production-ready design of magnetic sensors.

6.6 A multistorey three-dimensional steel structure: reliability analysis and optimisation

This section describes a case study in the field of structural design optimisation that focuses on a complex and challenging engineering problem. The study uses the PolyMoment-based RBDO method to evaluate uncertainty and ensure safety, by combining high-order moment-based uncertainty analysis with efficient response surface modelling. This method outperforms existing methods in terms of accuracy and computational efficiency, making it a significant case study in the field.

Figure 6.16 illustrates the finite element model of a five-story steel-framed modular building consisting of six identical corner-supported modules on each floor. The modules are 7.2 m long, 3.2 m wide, and 3.0 m high, with parallel-flange sections for beams and square hollow sections for columns. The floors are made of 100 mm thick concrete slabs, and all inter-module connections are assumed to be ideally pinned.

Uncertainties in steel-framed modular buildings often stem from the use of standardised prefabricated steel modules. When multiple modules are put together

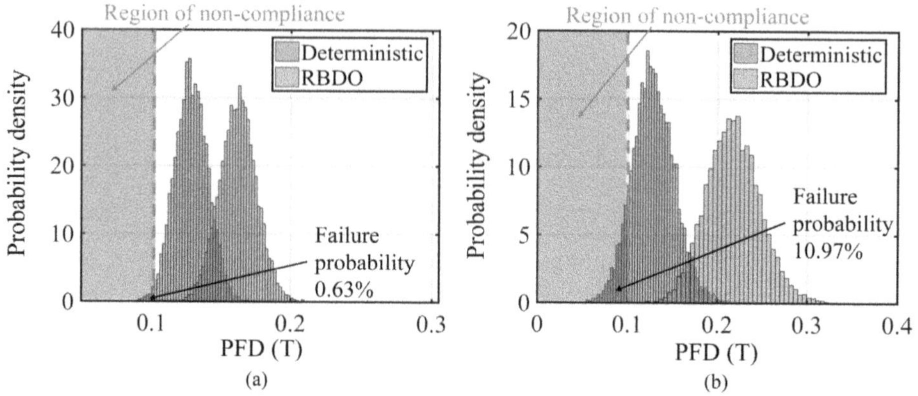

Figure 6.15. Probability distributions of the peak flux densities (PFDs) obtained using the MC method for deterministic and RBDO designs with manufacturing tolerances of (a) ±0.05 mm and (b) ±0.1 mm.

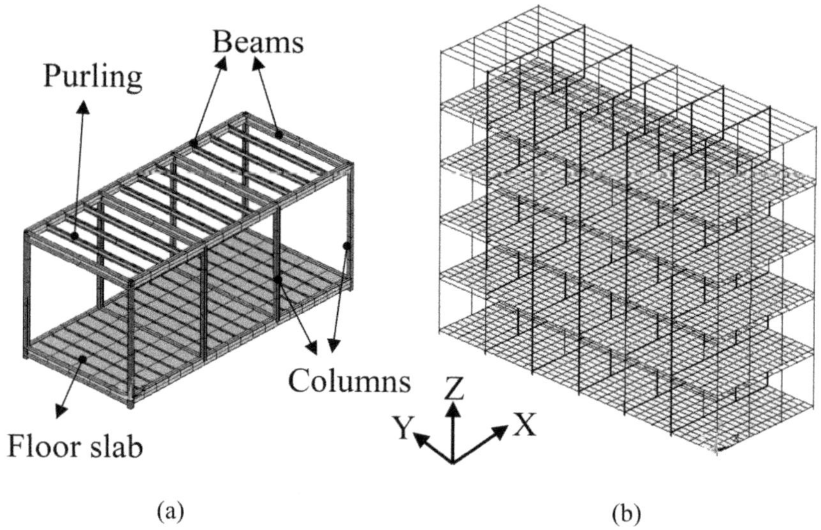

Figure 6.16. A three-dimensional multistorey structure in which (a) represents the single volumetric module and (b) represents the full structural model. Reprinted from [44], Copyright 2020, with permission from Elsevier.

to form a complex frame, it is crucial to consider the overall structural reliability. This is where the computational efficiency and accuracy of the PolyMoment method can be highly beneficial for engineers. The successful application of this method to a complex three-dimensional multistorey steel structure highlights its importance in understanding the implications of analytical uncertainty evaluation in engineering design.

6.6.1 Problem formulation

The FE model for the structure shown in figure 6.16 was obtained using the method outlined in [44]. The model represents a multistorey steel-framed structure that is subjected to a combination of loads, including self-weight and superimposed dead load, live load, and wind load. The dead load calculation was performed using the ANSYS gravity field [45]. The wind load was determined based on the wind speed for Cyclonic Region C as per the Standards Association of Australia's AS1170.2 [46], and the design wind direction was assumed to be parallel to the global X-direction.

Three performance functions were selected, including G_1, which is the inter-storey drift under SLS loading (AS1170.0 [3]), G_2, which is the ULS performance of columns subjected to combined axial N_x^* and flexural actions M_y^* and M_z^* (AS4100 [47]), and G_3, which is the ULS performance of beams (AS4100 [47]). The structure was considered to have failed if any one of the performance functions was not met. The target reliability index was set at $\beta = 3$ for G_1 and $\beta = 3.8$ for G_2 and G_3, as per ISO 13822 [48].

In the optimisation process, variables such as M_{cx} and M_{cy} represent the member moment capacities in the principal and minor axes, respectively. N_{cx} is the column capacity, Z_{be} and Z_{ce} are the section modules for beams and columns, respectively, A_{b} and A_{c} are the cross-sectional areas of beams and columns, and details such as the material and geometric properties can be found in table 6.5. The probabilistic models for load variables in this table were determined based on the Australian Building Codes Board Handbook [49] and the material models based on the Joint Committee on Structural Safety code [50].

The starting point for the optimisation was set at $\{w_{\mathrm{c}}, t_{\mathrm{c}}, w_{\mathrm{fb}}, d_{\mathrm{bb}}, t_{\mathrm{fb}}, t_{\mathrm{wb}}\} = \{100,\ 9.0,\ 133,\ 202,\ 7.0,\ 5.0\}$ mm. The optimisation of the structure was carried out by minimising the total volume of the columns and beams and reducing the structural self-weight, as described in detail below:

$$
\begin{aligned}
&\text{find:} && w_{\mathrm{c}},\ t_{\mathrm{c}},\ w_{\mathrm{fb}},\ d_{\mathrm{bb}},\ t_{\mathrm{fb}} \text{ and } t_{\mathrm{wb}}\\
&\text{minimise:} && f = \sum L_{\mathrm{c}}A_{\mathrm{c}} + \sum L_{\mathrm{b}}A_{\mathrm{b}}\\
&\text{subject to:} && G_1 = \Delta - \frac{H}{500} \leqslant 0\\
& && G_2 = \left(\frac{M_y^*}{M_{\mathrm{c}y}}\right)^{1.4} + \left(\frac{M_z^*}{M_{\mathrm{c}z}}\right)^{1.4} \leqslant 1,\\
& && \text{where } M_{\mathrm{c}i} = M_{\mathrm{cs}}\left(1 - \frac{N_x^*}{N_{\mathrm{c}x}}\right) \text{ for } i\\
& && = x,\ y,\ N_{\mathrm{c}x} = \alpha_{\mathrm{c}}A_{\mathrm{c}}f_y \text{ and}\\
& && M_{\mathrm{cs}} = Z_{\mathrm{ce}}f_y\\
& && G_3 = \alpha_{\mathrm{m}}\alpha_{\mathrm{s}}M_{\mathrm{bs}}, \text{ where } M_{\mathrm{bs}} = Z_{\mathrm{be}}f_y,\\
& && \text{where: } \alpha_{\mathrm{m}} = 1.0 H = 3\,\mathrm{m} \sum L_{\mathrm{c}} = 720 \sum L_{\mathrm{b}} = 1248.
\end{aligned}
\tag{6.10}
$$

Table 6.5. Properties of the random variables for the three-dimensional multistorey structure problem; CoV denotes the coefficient of variation.

	Description	Distribution	Median		CoV
			SLS	ULS	
$SIDL$	Superimposed dead load (Pa)	Lognormal	1000		0.10
LL	Live load (Pa)	Lognormal	1500		0.43
w_c	Width of column (Pa)	Normal	Not applicable		0.0205
t_c	Wall thickness of column (Pa)	Normal	Not applicable		0.0362
w_{fb}	Flange width of beam (Pa)	Normal	Not applicable		0.0132
d_{bb}	Depth of beam (Pa)	Normal	Not applicable		0.0364
t_{fb}	Flange thickness of beam (Pa)	Normal	Not applicable		0.0182
t_{wb}	Web thickness of beam (Pa)	Normal	Not applicable		0.0151
E_s	Elastic modulus of steel (Pa)	Lognormal	206×10^9		0.03
E_c	Elastic modulus of concrete (Pa)	Lognormal	21.8×10^9		0.15
p_W	Windward wind pressure (Pa)	Lognormal	169.5040	432.0960	0.16
p_L	Leeward wind pressure (Pa)	Lognormal	56.4987	144.0309	0.16
p_R	Roof wind pressure (Pa)	Lognormal	169.4961	432.0927	0.16
Rou_S	Density of steel (kg m^{-3})	Lognormal	7700		0.01
Rou_C	Density of concrete (kg m^{-3})	Lognormal	2400		0.04
f_y	Yield stress of steel (Pa)	Lognormal	350×10^6		0.05

6.6.2 The resultant design and benchmarking

Table 6.6 presents the results of the RBDO problem for the complex steel structure shown in figure 6.16; in this table, reliability analysis based on PMA from section 3.2 is utilised to compare the outcomes of the PolyMoment method, which was applied with up to $N_m = 4$ and $N_m = 8$ moments. The table reports the reliability indexes of the final RBDO designs, which were obtained from the ANSYS model of the structure. The reliability indexes were calculated using the Monte Carlo method. Only the final reliability index of performance function G_1 is shown in table 6.6, as the reliability indexes of both G_2 and G_3 are infinite. This implies that G_1 is the only constraint that plays a role in determining the final design.

The results reveal that the PolyMoment-based method requires significantly fewer function evaluations (540 evaluations) compared to the PMA method (6022 evaluations), leading to an approximately elevenfold increase in computational efficiency. The design obtained using the PolyMoment method meets the target reliability index of $\beta = 3$ as described in (6.10), while the design obtained using the MPP-based RBDO method does not meet this target, potentially resulting in an underdesigned structure.

In addition, the results in table 6.6 show that the use of higher-order moments improves the accuracy of the reliability analysis and results in a more optimum final design with a lower total cross-sectional area, without sacrificing the number of

Table 6.6. RBDO results for the three-dimensional multistorey structure problem. The reliability indexes β of G_2 and G_3 are infinite. The PMA was employed using a recent MPP-based algorithm provided in [51] for reliability analysis.

	Final design								
Method	w_c	t_c	w_{fb}	d_{bb}	t_{fb}	t_{wb}	C	FEV counts	β of G_1
Deterministic	0.0947	0.0050	0.0750	0.2300	0.0120	0.0060	5.0801	120	
PMA	0.0988	0.0050	0.0750	0.2300	0.0120	0.0060	5.1828	6022	2.8043
PolyMoment ($N_m = 4$)	0.0976	0.0050	0.0750	0.2300	0.0120	0.0060	5.2000	540	3.0274
PolyMoment ($N_m = 8$)	0.0968	0.0050	0.0750	0.2300	0.0120	0.0060	5.1108	540	3.0107

function evaluations. This efficiency and accuracy improvement is due to the absence of MPP iterations in finding the reliability constraints and the selective sampling paradigm in finding the local response surface models.

It is important to note that while the localised response surface with selective sampling mechanism could be used for MPP-based methods, the added assumptions on top of the MPP transformation and search strategies may negatively affect its reliability constraint evaluation accuracy. As such, the PolyMoment-based probabilistic optimisation methodology is a more desirable choice for complex problems, especially for safety-critical structural designs that are required to meet building standards and their reliability constraints, such as wind and earthquakes.

6.7 Summary

This chapter presented a unified moment-based probabilistic optimisation framework called the PolyMoment-based approach, which is generic and can be adapted to RDO, RBDO, and RBRDO optimisation paradigms. The methodology was explained in detail and its use was demonstrated through five real-world applications: lithium-ion batteries, vehicle side-impact crashworthiness, a magnetic sensor module, and a multistorey steel structure.

References

[1] Missoum S, Ramu P and Haftka R T 2007 A convex hull approach for the reliability-based design optimization of nonlinear transient dynamic problems *Comput. Meth. Appl. Mech. Eng.* **196** 2895–906

[2] Youn B D and Choi K K 2004 A new response surface methodology for reliability-based design optimization *Comput. Struct.* **82** 241–56

[3] 2002 *Structural design action—part 0: general principles* AS/NZS 1170.0:2002 Standards New Zealand https://www.standards.govt.nz/shop/asnzs-1170-02002

[4] Mansour R and Olsson M 2016 Response surface single loop reliability-based design optimization with higher-order reliability assessment *Struct. Multidiscip. Optim.* **54** 63–79

[5] Chen Z, Qiu H, Gao L, Li X and Li P 2014 A local adaptive sampling method for reliability-based design optimization using Kriging model *Struct. Multidiscip. Optim.* **49** 401–16

[6] Shang Y-W and Qiu Y-H 2006 A note on the extended Rosenbrock function *Evol. Comput.* **14** 119–26

[7] Rajan A, Vijayaraghavan V, Ooi M P-L, Garg A and Kuang Y C 2018 A simulation-based probabilistic framework for lithium-ion battery modelling *Measurement* **115** 87–94

[8] Abaqus version 6.10. *Abaqus User Subroutines Reference Manual*, Providence, RI Dassault Systèmes Simulia Corp., 2010

[9] Sahraei E, Kahn M, Meier J and Wierzbicki T 2015 Modelling of cracks developed in lithium-ion cells under mechanical loading *RSC Adv.* **5** 80369–80

[10] Dufo-López R, Lujano-Rojas J M and Bernal-Agustín J L 2014 Comparison of different lead–acid battery lifetime prediction models for use in simulation of stand-alone photovoltaic systems *Appl. Energy* **115** 242–53

[11] Yarime M, Shiroyama H and Kuroki Y 2008 The strategies of the Japanese auto industry in developing hybrid and fuel cell vehicles *Making Choices about Hydrogen: Transport Issues for Developing Countries* (Tokyo: UNU Press) pp 187–212 https://www.idrc.ca/en/book/making-choices-about-hydrogen-transport-issues-developing-countries International Development Research Centre (IDRC)

[12] Young K, Wang C, Wang L Y and Strunz K 2013 Electric vehicle battery technologies *Electric Vehicle Integration into Modern Power Networks* (New York: Springer) pp 15–56

[13] Iman R 2008 Latin Hypercube Sampling *Encyclopedia of Statistical Sciences* (Hoboken, NJ: Wiley)

[14] Draper N 1995 *Response Surface Methodology: Process and Product Optimization Using Designed Experiments: RH Myers and DC Montgomery* vol 59 (New York: Wiley) p 714

[15] Nocedal J and Wright S J 2006 Quadratic programming *Numerical Optimization* (New York: Springer) pp 448–92

[16] Deb K, Pratap A, Agarwal S and Meyarivan T 2002 A fast and elitist multiobjective genetic algorithm: NSGA-II *IEEE Trans. Evol. Comput.* **6** 182–97

[17] Tian Q, Zhao D, Li Z and Zhu Q 2017 Robust and stepwise optimization design for CO_2 pipeline transportation *Int. J. Greenhouse Gas Control* **58** 10–8

[18] Zhao D, Tian Q, Li Z and Zhu Q 2016 A new stepwise and piecewise optimization approach for CO_2 pipeline *Int. J. Greenhouse Gas Control* **49** 192–200

[19] Youn B D, Choi K, Yang R-J and Gu L 2004 Reliability-based design optimization for crashworthiness of vehicle side impact *Struct. Multidiscip. Optim.* **26** 272–83

[20] W. G. European Enhanced Vehicle-Safety Committee *Side Impact* https://eevc.net/EEVC/EN/Past/WG09/WG09.html (accessed January 2023)

[21] Rajan A, Ooi M, Kuang Y C and Demidenko S 2017 Reliability-based design optimisation of technical systems: analytical response surface moments method *J. Eng.* **2017** 36–46

[22] Valdebenito M A and Schuller G I 2010 A survey on approaches for reliability-based optimization *Struct. Multidiscip. Optim.* **42** 645–63

[23] Åhman M 2001 Primary energy efficiency of alternative powertrains in vehicles *Energy* **26** 973–89

[24] Schäfer A, Heywood J B and Weiss M A 2006 Future fuel cell and internal combustion engine automobile technologies: a 25-year life cycle and fleet impact assessment *Energy* **31** 2064–87

[25] Corbo P, Migliardini F and Veneri O 2007 Performance investigation of 2.4 kW PEM fuel cell stack in vehicles *Int. J. Hydrog. Energy* **32** 4340–9
[26] Srinivasan S, Velev O A, Parthasarathy A, Manko D J and Appleby A J 1991 High energy efficiency and high power density proton exchange membrane fuel cells—electrode kinetics and mass transport *J. Power Sources* **36** 299–320
[27] Verbrugge M W and Hill R F 1990 Ion and solvent transport in ion-exchange membranes: II. A radiotracer study of the sulfuric-acid, nation-117 system *J. Electrochem. Soc.* **137** 893
[28] Rajan A, Garg A, Vijayaraghavan V, Kuang Y C and Ooi M P-L 2018 Parameter optimization of polymer electrolyte membrane fuel cell using moment-based uncertainty evaluation technique *J. Energy Storage* **15** 8–16
[29] Xuan D, Li Z, Kim J and Kim Y 2009 Optimal operating points of PEM fuel cell model with RSM *J. Mech. Sci. Technol.* **23** 717–28
[30] Park C, Oh K, Kim D and Kim H 2004 Development of fuel cell hybrid electric vehicle performance simulator *Int. J. Automot. Technol.* **5** 287–95 http://www.ijat.net/journal/view.php?number=194
[31] Al-Baghdadi M A S and Al-Janabi H A S 2007 Parametric and optimization study of a PEM fuel cell performance using three-dimensional computational fluid dynamics model *Renew. Energy* **32** 1077–101
[32] Sikha G and Popov B N 2004 Performance optimization of a battery–capacitor hybrid system *J. Power Sources* **134** 130–8
[33] Kim M-J and Peng H 2007 Power management and design optimization of fuel cell/battery hybrid vehicles *J. Power Sources* **165** 819–32
[34] San F G B, Isik-Gulsac I and Okur O 2013 Analysis of the polymer composite bipolar plate properties on the performance of PEMFC (polymer electrolyte membrane fuel cells) by RSM (response surface methodology) *Energy* **55** 1067–75
[35] Okur O, Alper E and Almansoori A 2014 Optimization of catalyst preparation conditions for direct sodium borohydride fuel cell using response surface methodology *Energy* **67** 97–105
[36] Kanani H, Shams M, Hasheminasab M and Bozorgnezhad A 2015 Model development and optimization of operating conditions to maximize PEMFC performance by response surface methodology *Energy Convers. Manage.* **93** 9–22
[37] Charoen K *et al* 2017 Application of response surface methodology to optimize direct alcohol fuel cell power density for greener energy production *J. Clean. Prod.* **142** 1309–20
[38] Garg A, Panda B N and Lam J S L 2016 Functional characterization of current characteristic of direct methanol fuel cell *Fuel* **183** 432–40
[39] Fesanghary M, Mahdavi M, Minary-Jolandan M and Alizadeh Y 2008 Hybridizing harmony search algorithm with sequential quadratic programming for engineering optimization problems *Comput. Meth. Appl. Mech. Eng.* **197** 3080–91
[40] Tu J H, Luchtenburg D M, Rowley C W, Brunton S L and Kutz J N 2013 Novel sampling strategies for dynamic mode decomposition *66th Annual Meeting of the APS Division of Fluid Dynamics* 58 (College Park, MD: American Physical Society) APS Division of Fluid Dynamics Meeting Abstracts, H35.001 http://meetings.aps.org/link/BAPS.2013.DFD.H35.1
[41] Chatterjee A 2000 An introduction to the proper orthogonal decomposition *Curr. Sci.* **78** 808–17 https://www.currentscience.ac.in/Volumes/78/07/0808.pdf

[42] Paul S, Rajan A, Chang J, Kuang Y C and Ooi M P-L 2018 Parametric design analysis of magnetic sensor based on model order reduction and reliability-based design optimization *IEEE Trans. Magn.* **54** 1–4

[43] Paul S and Chang J 2017 Design and parametric study of the magnetic sensor for position detection in linear motor based on nonlinear parametric model order reduction *Sensors* **17** 1543

[44] Rajan A, Luo F J, Kuang Y C, Bai Y and Ooi M P-L 2020 Reliability-based design optimisation of structural systems using high-order analytical moments *Struct. Saf.* **86** 101970

[45] Stolarski T, Nakasone Y and Yoshimoto S 2018 *Engineering Analysis with ANSYS Software* 2 (Oxford: Butterworth-Heinemann)

[46] 2011 *Structural design actions - Part 2: Wind actions* AS/NZS 1170.2:2011 Standards New Zealand https://www.standards.govt.nz/shop/asnzs-1170-22011/

[47] 2016 *Steel structures* AS 4100—1998 Standards Australia https://www.standards.org.au/standards-catalogue/sa-snz/building/bd-001/as—4100-1998

[48] ISO 13822:2010 2010 *Bases for design of structures—Assessment of existing structures* (International Organisation for Standardization)

[49] 2019 *Handbook: Structural ReliabilityVerification Method* 1.0 (Canberra: Australian Building Codes Board)

[50] 2001 Probabilistic model code part 2: load models https://www.jcss-lc.org/jcss-probabilistic-model-code/ Joint Committee on Structural Safety https://www.jcss-lc.org/jcss-probabilistic-model-code/

[51] Roudak M A, Shayanfar M A, Barkhordari M A and Karamloo M 2017 A robust approximation method for nonlinear cases of structural reliability analysis *Int. J. Mech. Sci.* **133** 11–20

IOP Publishing

Analytical Evaluation of Uncertainty Propagation for Probabilistic Design Optimisation

Melanie Po-Leen Ooi, Arvind Rajan, Ye Chow Kuang and Serge Demidenko

Appendix A

Lookup table for the parameters *g* and *h* in Tukey's *gh* distribution

The entries for the values of *g* and *h* in the table below satisfy their boundary conditions and were obtained using a genetic algorithm in MATLAB [1]. The symbol $|\,\Omega\,|$ represents absolute skewness and $K - 3$ represents excess kurtosis.

doi:10.1088/978-0-7503-4931-4ch7

|Ω|	K − 3	g	h	|Ω|	K − 3	g	h	|Ω|	K − 3	g	h	|Ω|	K − 3	g	h
0.00	0.0	0.0000	0.0000	0.5	0.5	0.0921	0.1063	0.25	1.2	0.0411	0.1233	0.1	2.1	0.0151	0.1344
0.05	0.0	0.0091	0.1088	0.55	0.5	0.1026	0.1040	0.3	1.2	0.0496	0.1224	0.15	2.1	0.0227	0.1340
0.10	0.0	0.0183	0.1085	0.6	0.5	0.1137	0.1015	0.35	1.2	0.0583	0.1212	0.2	2.1	0.0304	0.1335
0.15	0.0	0.0275	0.1079	0.65	0.5	0.1251	0.0986	0.4	1.2	0.0672	0.1199	0.25	2.1	0.0382	0.1329
0.20	0.0	0.0370	0.1071	0.7	0.5	0.1372	0.0954	0.45	1.2	0.0763	0.1183	0.3	2.1	0.0461	0.1320
0.25	0.0	0.0464	0.1061	0	0.6	0.0000	0.0000	0.5	1.2	0.0858	0.1166	0.35	2.1	0.0541	0.1310
0.30	0.0	0.0561	0.1049	0.05	0.6	0.0086	0.1178	0.55	1.2	0.0954	0.1146	0	2.2	0.0000	0.0000
0.35	0.0	0.0660	0.1034	0.1	0.6	0.0172	0.1176	0	1.3	0.0000	0.0000	0.05	2.2	0.0075	0.1355
0.40	0.0	0.0762	0.1017	0.15	0.6	0.0258	0.1171	0.05	1.3	0.0080	0.1265	0.1	2.2	0.0150	0.1353
0.45	0.0	0.0868	0.0997	0.2	0.6	0.0346	0.1164	0.1	1.3	0.0161	0.1263	0.15	2.2	0.0226	0.1350
0.50	0.0	0.0976	0.0974	0.25	0.6	0.0435	0.1155	0.15	1.3	0.0242	0.1258	0.2	2.2	0.0302	0.1344
0.55	0.0	0.1090	0.0948	0.3	0.6	0.0525	0.1144	0.2	1.3	0.0324	0.1253	0.25	2.2	0.0379	0.1338
0.60	0.0	0.1209	0.0919	0.35	0.6	0.0618	0.1131	0.25	1.3	0.0407	0.1245	0.3	2.2	0.0458	0.1330
0.65	0.0	0.1333	0.0887	0.4	0.6	0.0712	0.1116	0.3	1.3	0.0491	0.1236	0.35	2.2	0.0537	0.1320
0.70	0.0	0.1463	0.0851	0.45	0.6	0.0810	0.1099	0.35	1.3	0.0578	0.1224	0	2.3	0.0000	0.0000
0.75	0.0	0.1602	0.0810	0.5	0.6	0.0910	0.1079	0.4	1.3	0.0666	0.1211	0.05	2.3	0.0074	0.1364
0.80	0.0	0.1748	0.0766	0.55	0.6	0.1015	0.1057	0.45	1.3	0.0756	0.1196	0.1	2.3	0.0149	0.1362
0.00	0.1	0.0000	0.0000	0.6	0.6	0.1124	0.1032	0.5	1.3	0.0850	0.1179	0.15	2.3	0.0224	0.1359
0.05	0.1	0.0090	0.1104	0.65	0.6	0.1238	0.1004	0	1.4	0.0000	0.0000	0.2	2.3	0.0300	0.1354
0.10	0.1	0.0181	0.1101	0	0.7	0.0000	0.0000	0.05	1.4	0.0080	0.1276	0.25	2.3	0.0376	0.1347
0.15	0.1	0.0272	0.1096	0.05	0.7	0.0085	0.1192	0.1	1.4	0.0160	0.1274	0.3	2.3	0.0454	0.1339
0.20	0.1	0.0365	0.1088	0.1	0.7	0.0170	0.1189	0.15	1.4	0.0240	0.1270	0	2.4	0.0000	0.0000
0.25	0.1	0.0459	0.1078	0.15	0.7	0.0256	0.1184	0.2	1.4	0.0321	0.1264	0.05	2.4	0.0074	0.1373
0.30	0.1	0.0554	0.1066	0.2	0.7	0.0342	0.1178	0.25	1.4	0.0404	0.1256	0.1	2.4	0.0148	0.1371
0.35	0.1	0.0652	0.1052	0.25	0.7	0.0430	0.1169	0.3	1.4	0.0487	0.1247	0.15	2.4	0.0222	0.1367
0.40	0.1	0.0753	0.1035	0.3	0.7	0.0520	0.1159	0.35	1.4	0.0573	0.1236	0.2	2.4	0.0297	0.1362
0.45	0.1	0.0857	0.1015	0.35	0.7	0.0612	0.1146	0.4	1.4	0.0660	0.1223	0.25	2.4	0.0374	0.1356
0.50	0.1	0.0964	0.0993	0.4	0.7	0.0705	0.1131	0.45	1.4	0.0750	0.1208	0.3	2.4	0.0451	0.1348
0.55	0.1	0.1077	0.0968	0.45	0.7	0.0801	0.1114	0.5	1.4	0.0842	0.1191	0	2.5	0.0000	0.0000

(Continued)

0.60	0.1	0.1193	0.0940	0.5	0.7	0.0901	0.1095	0	1.5	0.0000	0.0000	0.05	2.5	0.0073	0.1381
0.65	0.1	0.1316	0.0908	0.55	0.7	0.1004	0.1073	0.05	1.5	0.0079	0.1287	0.1	2.5	0.0147	0.1379
0.70	0.1	0.1444	0.0873	0.6	0.7	0.1112	0.1048	0.1	1.5	0.0158	0.1285	0.15	2.5	0.0221	0.1376
0.75	0.1	0.1580	0.0834	0.65	0.7	0.1223	0.1021	0.15	1.5	0.0238	0.1281	0.2	2.5	0.0296	0.1371
0.00	0.2	0.0000	0.0000	0	0.8	0.0000	0.0000	0.2	1.5	0.0319	0.1275	0.25	2.5	0.0371	0.1365
0.05	0.2	0.0089	0.1120	0.05	0.8	0.0084	0.1205	0.25	1.5	0.0400	0.1267	0.3	2.5	0.0447	0.1357
0.10	0.2	0.0179	0.1117	0.1	0.8	0.0168	0.1202	0.3	1.5	0.0482	0.1258	0	2.6	0.0000	0.0000
0.15	0.2	0.0269	0.1112	0.15	0.8	0.0253	0.1198	0.35	1.5	0.0568	0.1247	0.05	2.6	0.0073	0.1390
0.20	0.2	0.0361	0.1104	0.2	0.8	0.0339	0.1191	0.4	1.5	0.0654	0.1235	0.1	2.6	0.0146	0.1388
0.25	0.2	0.0454	0.1095	0.25	0.8	0.0426	0.1183	0.45	1.5	0.0743	0.1220	0.15	2.6	0.0219	0.1384
0.30	0.2	0.0548	0.1083	0.3	0.8	0.0515	0.1172	0.5	1.5	0.0834	0.1204	0.2	2.6	0.0293	0.1379
0.35	0.2	0.0645	0.1069	0.35	0.8	0.0605	0.1160	0	1.6	0.0000	0.0000	0.25	2.6	0.0369	0.1373
0.40	0.2	0.0745	0.1052	0.4	0.8	0.0698	0.1145	0.05	1.6	0.0078	0.1298	0	2.7	0.0000	0.0000
0.45	0.2	0.0847	0.1033	0.45	0.8	0.0793	0.1129	0.1	1.6	0.0157	0.1295	0.05	2.7	0.0072	0.1398
0.50	0.2	0.0953	0.1011	0.5	0.8	0.0892	0.1110	0.15	1.6	0.0236	0.1291	0.1	2.7	0.0145	0.1396
0.55	0.2	0.1063	0.0987	0.55	0.8	0.0993	0.1089	0.2	1.6	0.0316	0.1286	0.15	2.7	0.0218	0.1392
0.60	0.2	0.1179	0.0959	0.6	0.8	0.1099	0.1065	0.25	1.6	0.0397	0.1278	0.2	2.7	0.0292	0.1388
0.65	0.2	0.1298	0.0929	0	0.9	0.0000	0.0000	0.3	1.6	0.0480	0.1269	0.25	2.7	0.0366	0.1382
0.70	0.2	0.1425	0.0894	0.05	0.9	0.0083	0.1218	0.35	1.6	0.0563	0.1259	0	2.8	0.0000	0.0000
0.75	0.2	0.1558	0.0857	0.1	0.9	0.0167	0.1215	0.4	1.6	0.0649	0.1246	0.05	2.8	0.0072	0.1406
0.00	0.3	0.0000	0.0000	0.15	0.9	0.0251	0.1211	0.45	1.6	0.0737	0.1232	0.1	2.8	0.0144	0.1404
0.05	0.3	0.0088	0.1135	0.2	0.9	0.0336	0.1204	0	1.7	0.0000	0.0000	0.15	2.8	0.0217	0.1400
0.10	0.3	0.0177	0.1132	0.25	0.9	0.0422	0.1196	0.05	1.7	0.0078	0.1308	0.2	2.8	0.0290	0.1396
0.15	0.3	0.0266	0.1127	0.3	0.9	0.0510	0.1186	0.1	1.7	0.0156	0.1305	0	2.9	0.0000	0.0000
0.20	0.3	0.0357	0.1120	0.35	0.9	0.0599	0.1174	0.15	1.7	0.0234	0.1302	0.05	2.9	0.0071	0.1413
0.25	0.3	0.0448	0.1111	0.4	0.9	0.0691	0.1159	0.2	1.7	0.0314	0.1296	0.1	2.9	0.0143	0.1412
0.30	0.3	0.0542	0.1099	0.45	0.9	0.0785	0.1143	0.25	1.7	0.0394	0.1289	0.15	2.9	0.0215	0.1408
0.35	0.3	0.0637	0.1085	0.5	0.9	0.0882	0.1125	0.3	1.7	0.0475	0.1280	0.2	2.9	0.0288	0.1404
0.40	0.3	0.0736	0.1069	0.55	0.9	0.0983	0.1104	0.35	1.7	0.0558	0.1270	0	3	0.0000	0.0000
0.45	0.3	0.0837	0.1050	0.6	0.9	0.1088	0.1080	0.4	1.7	0.0644	0.1257	0.05	3	0.0071	0.1421
0.50	0.3	0.0942	0.1029	0	1	0.0000	0.0000	0.45	1.7	0.0731	0.1243	0.1	3	0.0142	0.1419

(Continued)

(Continued)

$\lvert\Omega\rvert$	$K-3$	g	h	$\lvert\Omega\rvert$	$K-3$	g	h	$\lvert\Omega\rvert$	$K-3$	g	h	$\lvert\Omega\rvert$	$K-3$	g	h
0.55	0.3	0.1051	0.1005	0.05	1	0.0082	0.1230	0	1.8	0.0000	0.0000	0.15	3	0.0214	0.1416
0.60	0.3	0.1164	0.0978	0.1	1	0.0165	0.1227	0.05	1.8	0.0077	0.1318	0.2	3	0.0286	0.1411
0.65	0.3	0.1282	0.0949	0.15	1	0.0248	0.1223	0.1	1.8	0.0154	0.1316	0	3.1	0.0000	0.0000
0.70	0.3	0.1407	0.0915	0.2	1	0.0333	0.1217	0.15	1.8	0.0233	0.1312	0.05	3.1	0.0071	0.1428
0.00	0.4	0.0000	0.0000	0.25	1	0.0419	0.1209	0.2	1.8	0.0311	0.1306	0.1	3.1	0.0141	0.1427
0.05	0.4	0.0087	0.1150	0.3	1	0.0505	0.1199	0.25	1.8	0.0391	0.1299	0.15	3.1	0.0213	0.1423
0.10	0.4	0.0175	0.1147	0.35	1	0.0593	0.1187	0.3	1.8	0.0471	0.1291	0	3.2	0.0000	0.0000
0.15	0.4	0.0264	0.1142	0.4	1	0.0685	0.1173	0.35	1.8	0.0554	0.1280	0.05	3.2	0.0070	0.1436
0.20	0.4	0.0353	0.1135	0.45	1	0.0778	0.1157	0.4	1.8	0.0638	0.1268	0.1	3.2	0.0140	0.1434
0.25	0.4	0.0444	0.1126	0.5	1	0.0874	0.1139	0	1.9	0.0000	0.0000	0.15	3.2	0.0211	0.1431
0.30	0.4	0.0536	0.1115	0.55	1	0.0974	0.1118	0.05	1.9	0.0076	0.1328	0	3.3	0.0000	0.0000
0.35	0.4	0.0630	0.1101	0.6	1	0.1077	0.1095	0.1	1.9	0.0153	0.1325	0.05	3.3	0.0070	0.1443
0.40	0.4	0.0728	0.1085	0	1.1	0.0000	0.0000	0.15	1.9	0.0231	0.1321	0.1	3.3	0.0140	0.1441
0.45	0.4	0.0828	0.1067	0.05	1.1	0.0082	0.1242	0.2	1.9	0.0309	0.1316	0	3.4	0.0000	0.0000
0.50	0.4	0.0931	0.1047	0.1	1.1	0.0164	0.1240	0.25	1.9	0.0388	0.1309	0.05	3.4	0.0069	0.1450
0.55	0.4	0.1039	0.1023	0.15	1.1	0.0247	0.1235	0.3	1.9	0.0468	0.1301	0.1	3.4	0.0139	0.1448
0.60	0.4	0.1150	0.0997	0.2	1.1	0.0330	0.1229	0.35	1.9	0.0550	0.1291	0	3.5	0.0000	0.0000
0.65	0.4	0.1266	0.0968	0.25	1.1	0.0414	0.1221	0.4	1.9	0.0633	0.1279	0.05	3.5	0.0069	0.1457
0.70	0.4	0.1389	0.0935	0.3	1.1	0.0500	0.1211	0	2	0.0000	0.0000	0.1	3.5	0.0138	0.1455
0.00	0.5	0.0000	0.0000	0.35	1.1	0.0588	0.1200	0.05	2	0.0076	0.1337	0	3.6	0.0000	0.0000
0.05	0.5	0.0086	0.1165	0.4	1.1	0.0679	0.1186	0.1	2	0.0152	0.1335	0.05	3.6	0.0069	0.1464
0.10	0.5	0.0173	0.1162	0.45	1.1	0.0770	0.1170	0.15	2	0.0229	0.1331	0	3.7	0.0000	0.0000
0.15	0.5	0.0261	0.1157	0.5	1.1	0.0866	0.1152	0.2	2	0.0306	0.1326	0.05	3.7	0.0068	0.1470
0.20	0.5	0.0349	0.1150	0.55	1.1	0.0964	0.1132	0.25	2	0.0385	0.1319	0	3.8	0.0000	0.0000
0.25	0.5	0.0439	0.1141	0	1.2	0.0000	0.0000	0.3	2	0.0464	0.1311	0	3.9	0.0000	0.0000
0.30	0.5	0.0531	0.1130	0.05	1.2	0.0081	0.1254	0.35	2	0.0545	0.1301	0	4	0.0000	0.0000
0.35	0.5	0.0624	0.1117	0.1	1.2	0.0162	0.1251	0.4	2	0.0628	0.1289				
0.40	0.5	0.0720	0.1101	0.15	1.2	0.0244	0.1247	0	2.1	0.0000	0.0000				
0.45	0.5	0.0819	0.1083	0.2	1.2	0.0327	0.1241	0.05	2.1	0.0075	0.1346				

Reference

[1] MATLAB version 9.1.0.441655 (R2016b). Natick, Massachusetts: The MathWorks Inc., 2016

Appendix B

Lookup table for the Mellin transforms of various families of probability distribution

This appendix contains the Mellin transforms of commonly used probability distributions. Note that $\delta, \omega, \omega_1, \omega_2 > 0$ and $-\infty < v < \infty$, $(a)_n = a(a-1)\cdots(a-n+1)$ follow the Pochhammer notation [1], $a!!$ denotes the double factorial, and $\mathcal{M}\left[f_{V^m}^{+}(v)\right]$ denotes the Mellin transform of the probability density function $f_{V^m}^{+}(v)$, whereby V is a random variable with a standard tabulated distribution and m is an integer. Column one defines the positive half of the PDF $f_V^{+}(v)$ of V, where ω and δ represent the shape and scale parameters, respectively. Column two lists the corresponding Mellin transform $\mathcal{M}\left[f_V^{+}(v)\right](s)$ for each distribution $f_V^{+}(v)$. Only one-sided (positive) transforms are listed, because this tabulation enables the flexible evaluation of one-sided, two-sided symmetrical, or two-sided asymmetrical distributions. The appropriate usage will be discussed in the following subsections. Columns three and four list the simplified expressions for the Mellin transform of $f_V^{+}(v)$, denoted by $M_V^{+}(s) = \mathcal{M}\left[f_V^{+}(v)\right](s)$, which is crucial in determining $\mathrm{E}[V^m]$ [2].

doi:10.1088/978-0-7503-4931-4ch8

Distribution	$M_V^+(s) = \mathcal{M}\left[f_V^+(v)\right](s)$	Simplified form of $M_V^+(m+1)$ for $m \in \mathbb{Z}^+$	
		For odd m	For even m
Normal: $f(v) = \frac{1}{\delta\sqrt{2\pi}} e^{-\frac{v^2}{2\delta^2}}$	$\delta^{s-1}\left[\frac{2^{(s-3)/2}\Gamma\left(\frac{s}{2}\right)}{\sqrt{\pi}}\right]$	$\delta^m\left[\frac{2^{m/2-1}\left(\frac{m-1}{2}\right)!}{\sqrt{\pi}}\right]$	$\delta^m\left[\frac{(m-1)!!}{2}\right]$
Student's t: $f(v) = \frac{\Gamma\left(\frac{\omega+1}{2}\right)}{2\delta\sqrt{\omega\pi}\,\Gamma\left(\frac{\omega}{2}\right)}\left(1 + \frac{1}{\omega}\left(\frac{v}{\delta}\right)^2\right)^{-\frac{\omega+1}{2}}$	$\delta^{s-1}\left[\frac{\omega^{\frac{(s-1)}{2}}\Gamma\left(\frac{s}{2}\right)\Gamma\left(\frac{\omega+1}{2}-\frac{s}{2}\right)}{2\sqrt{\pi}\,\Gamma\left(\frac{\omega}{2}\right)}\right]$	$\delta^{s-1}\left[\frac{\omega^{\frac{m}{2}}\Gamma\left(\frac{m+1}{2}\right)\Gamma\left(\frac{\omega-m}{2}\right)}{2\sqrt{\pi}\,\Gamma\left(\frac{\omega}{2}\right)}\right]$	$\delta^{s-1}\left[\frac{\omega^{\frac{m}{2}}\Gamma\left(\frac{m+1}{2}\right)\Gamma\left(\frac{\omega-m}{2}\right)}{2\sqrt{\pi}\,\Gamma\left(\frac{\omega}{2}\right)}\right]$
Triangular: $f(v) = \begin{cases} \frac{1}{\delta}\left(1 - \frac{\lvert v\rvert}{\delta}\right) \text{if } \lvert v\rvert \leqslant \delta \\ 0 \text{ otherwise} \end{cases}$	$\delta^{s-1}\left[\frac{\Gamma(s)}{\Gamma(2+s)}\right]$	$\delta^m\left[\frac{1}{m^2+3m+2}\right]$	$\delta^m\left[\frac{1}{m^2+3m+2}\right]$
Trapezoidal: $f(v) = \begin{cases} \frac{1}{\delta(1+\omega)} \text{ if } \lvert v\rvert \leqslant \omega \\ \frac{\delta-\lvert v\rvert}{\delta^2(1-\omega^2)} \text{ if } \omega < \lvert v\rvert \leqslant \delta \\ 0 \text{ otherwise} \end{cases}$	$\delta^{s-1}\left[\frac{(1-\omega^{s+1})}{(1-\omega^2)}\frac{\Gamma(s)}{\Gamma(2+s)}\right]$	$\delta^m\left[\frac{\frac{(1-\omega^{m+2})}{(1-\omega^2)}}{m^2+3m+2}\right]$	$\delta^m\left[\frac{\frac{(1-\omega^{m+2})}{(1-\omega^2)}}{m^2+3m+2}\right]$
Beta: $f(v) = \frac{1}{2\delta B(\omega_1,\omega_2)}\left(\frac{\lvert v\rvert}{\delta}\right)^{\omega_1-1}\left(1 - \frac{\lvert v\rvert}{\delta}\right)^{\omega_2-1}$	$\delta^{s-1}\left[\frac{\Gamma(\omega_1+\omega_2)\Gamma(\omega_1-1+s)}{2\Gamma(\omega_1)\Gamma(\omega_1+\omega_2-1+s)}\right]$	$\delta^m\left[\frac{(\omega_1+m-1)_m}{2(\omega_1+\omega_2+m-1)_m}\right]$	$\delta^m\left[\frac{(\omega_1+m-1)_m}{2(\omega_1+\omega_2+m-1)_m}\right]$
Uniform: $f(v) = \frac{1}{2\delta}$	$\delta^{s-1}\left[\frac{\Gamma(s)}{2\Gamma(1+s)}\right]$	$\delta^m\left[\frac{\frac{1}{2}}{m+1}\right]$	$\delta^m\left[\frac{\frac{1}{2}}{m+1}\right]$

Gamma: $f(v) = \frac{1}{2\delta\Gamma(\omega)}\left(\frac{\lvert v \rvert}{\delta}\right)^{\omega-1} e^{-\frac{\lvert v \rvert}{\delta}}$	$\delta^{s-1}\left[\frac{\Gamma(\omega-1+s)}{2\Gamma(\omega)}\right]$	$\delta^{m}\left[\frac{(\omega+m-1)_m}{2}\right]$	$\delta^{m}\left[\frac{(\omega+m-1)_m}{2}\right]$
Laplace: $f(v) = \frac{1}{2\delta}e^{-\frac{\lvert v \rvert}{\sigma}}$	$\delta^{s-1}\left[\frac{\Gamma(s)}{2}\right]$	$\delta^{m}\left[\frac{m!}{2}\right]$	$\delta^{m}\left[\frac{m!}{2}\right]$
Weibull: $f(v) = \frac{\omega}{\delta}\left(\frac{\lvert v \rvert}{\delta}\right)^{\omega-1} e^{-\left(\frac{\lvert v \rvert}{\delta}\right)^{\omega}}$	$\delta^{s-1}\left[\frac{\omega}{2}\Gamma\left(1-\frac{1}{\omega}+\frac{s}{\omega}\right)\right]$	$\delta^{m}\left[\frac{\omega\Gamma\left(\frac{m}{\omega}+1\right)}{2}\right]$	$\delta^{m}\left[\frac{\omega\Gamma\left(\frac{m}{\omega}+1\right)}{2}\right]$
Maxwell: $f(v) = \frac{1}{\delta}\sqrt{\frac{2}{\pi}}\left(\frac{v}{\delta}\right)^{2} e^{-\frac{v^2}{2\delta^2}}$	$\delta^{s-1}\left[\frac{2^{\frac{(s+1)}{2}}}{\sqrt{\pi}}\Gamma\left(1+\frac{s}{2}\right)\right]$	$\delta^{m}\left[\frac{2^{\frac{m}{2}+1}}{\sqrt{\pi}}\left(\frac{m+1}{2}\right)!\right]$	$\delta^{m}[(m+1)!!]$
Lognormal: $f(v) = \frac{1}{v\delta\sqrt{2\pi}}e^{-\frac{\ln^2(v)}{2\delta^2}}$	$\exp\left(\frac{(s-1)^2\delta^2}{2}\right)$	$\exp\left(\frac{m^2\delta^2}{2}\right)$	$\exp\left(\frac{m^2\delta^2}{2}\right)$

References

[1] MATLAB version 9.1.0.441655 (R2016b). Natick, Massachusetts: The MathWorks Inc., 2016

[2] Weisstein E W 2002 *CRC Concise Encyclopedia of Mathematics* (Boca Raton, FL: CRC Press)

www.ingramcontent.com/pod-product-compliance
Ingram Content Group UK Ltd.
Pitfield, Milton Keynes, MK11 3LW, UK
UKHW052041301026
9729IPUK00019B/358

9 780750 349321